Transducers in measurement and control

PETER H. SYDENHAM

M.E., Ph.D., F.Inst.M.C., F.I.I.C.A.

School of Electronic Engineering, South Australian Institute of Technology

Adam Hilger Ltd, Bristol

British Library Cataloguing in Publication Data
Sydenham, Peter H.
Transducers in measurement and control. — 2nd ed.
1. Transducers
I. Title
621.37'9 TJ223.T7

ISBN 0-85274-401-3

First published as a series of articles in *ELECTRONICS TODAY INTERNATIONAL* 1972/73.
Articles collected together and published in one volume by The University of New England Publishing Unit, Armidale, Australia, 1975.

This revised edition published by Adam Hilger Ltd, Techno House, Redcliffe Way, Bristol, BS1 6NX.
The Adam Hilger book-publishing imprint is owned by The Institute of Physics.

Printed and bound in Great Britain by J. H. Haynes & Co. Ltd, Sparkford, Yeovil, Somerset, BA22 7JJ, England.

Transducers in measurement and control

Contents

Preface

Transducers are low-powered machines devised by man to convert variables of the physical world into signals more suitable for the purposes of measurement and control.

Originally published in 1972/73 as a series of articles in the Australian edition of the international monthly *ELECTRONICS TODAY INTERNATIONAL* and, subsequently, as a combined volume by the University of New England Publishing Unit, New South Wales, Australia, this book provides a broad up-to-the-minute summary of the most commonly used transducers vital to our technological age.

It is intended to supplement, rather than to replace, existing books on transducers, containing, as it does, information on advanced methods that have found application in recent years and that, in general, have not previously been discussed in other works on transducers.

It contains chapters on large-scale length, position, tilt and alignment and on pollution measurement, all subjects not generally found in the transducer literature.

This second edition has been extensively revised to bring the material presented up to date. Every chapter now has a 'further reading' guide that includes the latest of published works. Several illustrations have been replaced to show the most recent version of the equipment reported and, in some instances, to provide a clearer illustration.

I am indebted to Collyn Rivers, Editorial Director of *ELECTRONICS TODAY INTERNATIONAL*, for the attention he gave to editing and producing the original series of articles. It was his suggestion that originated this successful work.

I am most pleased to acknowledge the many people who have supplied me with data, reprints, illustrations and, often, their invaluable time. They are, however, too numerous to mention personally.

I am grateful to all firms and institutions who have allowed me to reproduce information and illustrations and trust that the few firms with whom I was unable to make contact will excuse my presumption of their permission to use their material.

Finally, I wish to thank Jean Clinch and Coleen Drummond who prepared the typed manuscripts, Ron Farley who converted my original material into expert illustrations, Merv Bone who assisted with revision of the figures for the second edition, Ann Burke who converted the original articles into the first comprehensive book form and, finally, to the staff of Adam Hilger Ltd for their work on revising and restructuring the material to my broad directions.

Armidale, NSW, Australia
November 1978

Peter H. Sydenham

Acknowledgments for Illustrations

Acknowledgments, with appreciation, are due to the following sources, who supplied the illustrations used in this work. Inclusion of these pictures does not, however, carry any endorsement of the products featured. They have been selected to illustrate principles of measurement and the practice that has been, or is currently being, used to implement them.

Pioden Controls, Canterbury, England
Krautkramer GmbH, Cologne, West Germany
Hewlett-Packard Corp., Palo Alto, U.S.A.
Société Genevoise d'Instruments de Physique, Geneva, Switzerland
N.V. Philips Gloeilampenfabrieken, Eindhoven, The Netherlands
Innocenti Santeustacchio, Milan, Italy
Muirhead Ltd, Morden, England
Ferranti Ltd, Dalkieth, Scotland
Whitwell Electronic Developments, Glasgow, Scotland
Dr G. Manzoni, University of Trieste, Italy
Siemens AG, Berlin, West Germany
Moore Special Tool Co. Inc., Bridgepoint, U.S.A.
Data Technology Inc., Watertown, U.S.A.
Computer Control Co. Inc., Framington, U.S.A.
Laser Electronics Pty Ltd, Southport, Australia
ANAC Ltd, Auckland, New Zealand
British Aerospace Dynamics Group, Stevenage, England
Rank Taylor Hobson, Leicester, England
VEB Carl Zeiss JENA, Jena, German Democratic Republic
Dynasciences Corp., Blue Bell, U.S.A.
Electro-optics Associates, Palo Alto, U.S.A.
Grumman Aircraft Engineering Corp., U.S.A.
Schiess AG, Dusseldorf, West Germany
Bendix Automation and Measurement Division, Dayton, U.S.A.
Honeywell Pty Ltd, Sydney, Australia
IBES Pty Ltd, Adelaide, Australia
BOC Ltd, London, England
Elsinger-Feinmechanik, Zurich, Switzerland
The Military Engineer, New York, U.S.A.
Wild Heerbrugg, Heerbrugg, Switzerland
Cambridge Scientific Instruments Ltd, Cambridge, England
CSIRO, Melbourne, Australia
P. M. Tamson N.V., The Hague, The Netherlands
Leeds & Northrup Co., North Wales, U.S.A.
Institute of Physics, Bristol, England
Daedalus Enterprises Inc., Ann Arbor, U.S.A.
Ambac Industries, Pittsburgh, U.S.A.
Yellow Springs Instrument Co. Inc., Yellow Springs, U.S.A.

Wireless World, London, England
EG&G International Inc., Environmental Equipment Division, Waltham, U.S.A.
Institute of Measurement and Control, London, England
McGraw-Hill Book Co., New York, U.S.A.
Solartron Electronic Group, Farnborough, England
National Bureau of Standards, Washington D.C., U.S.A.
Ventron Corp., Cahn Division, Cerritos, U.S.A.
Martin Decker Corp., Santa Anna, U.S.A.
Mensor Corp., San Marcos, U.S.A.
Dawe Instruments Ltd, London, England
Travenol Laboratories Inc., Silver Springs, U.S.A.
Varian-Techtron Pty Ltd, Springvale, Australia
Varian Associates, Palo Alto, U.S.A.
Weathermeasure Corp., Sacramento, U.S.A.
Casella and Co. Ltd, London, England
Sigrist-Photometer AG, Zurich, Switzerland
Corning Medical, Halstead, England
George Kent Ltd, Sydney, Australia
Staff of *ELECTRONICS TODAY INTERNATIONAL*, Sydney, Australia, who prepared the majority of the line diagrams

CHAPTER 1

GENERAL PRINCIPLES, MICRODISPLACEMENT SENSORS

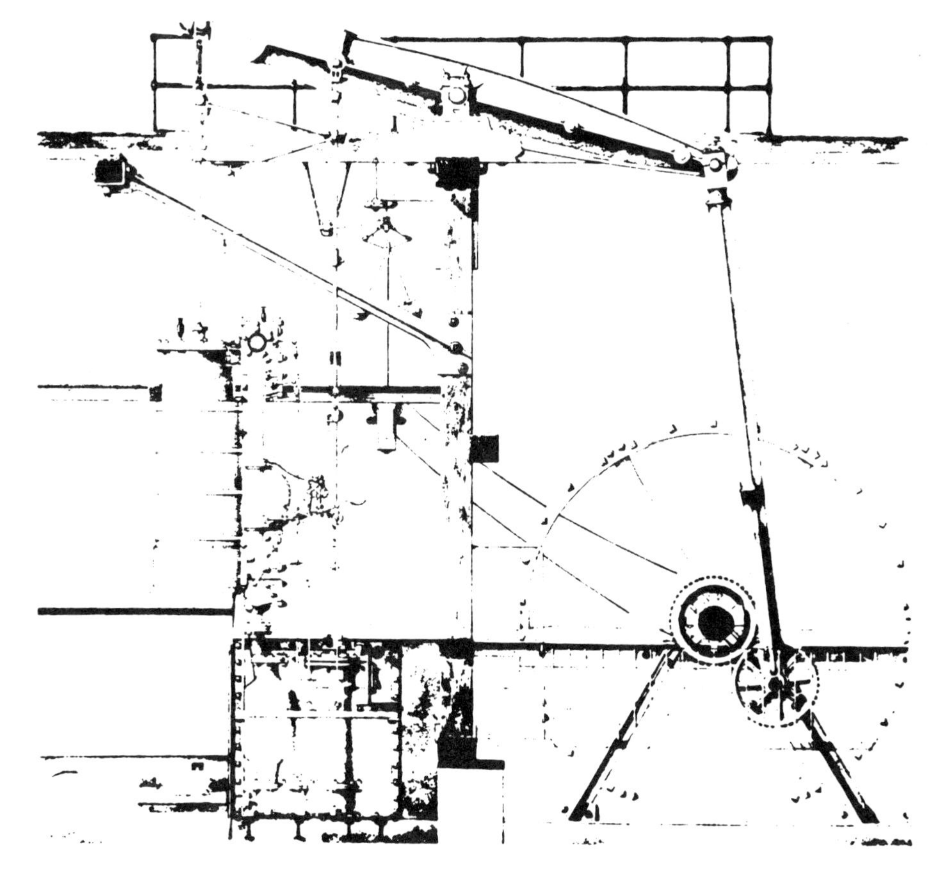

Boulton and Watt steam engine, built in 1784. A centrifugal ball speed-control governor is driven by a belt from the output wheel.

A TRANSDUCER is a device that converts (transduces) one physical variable into another. Transducers are not restricted to electrical signal conversion techniques, but in the main these predominate as electrical methods are universal, and provide a common interconnecting method for an engineering system or a scientific experiment.

This book will describe the proven practical methods (and this includes economic sense, as cost is important) now used to produce, in the main, electrical signals from the original physical effect to be measured.

Transducers provide convenient signals for measuring a process, for automatically recording these measurements when needed and, finally, for providing a signal that can be used to control. It is not possible to control without measuring and so the fundamental basis of automation is the transducer. The transducer is also able to provide gain by amplifying weak original signals before they are used. Amplification factors of a million are commonplace.

Often, more than one basic transducer principle is used to produce the required output. Units are cascaded. Consider the fuel gauge of a motor car, shown diagrammatically in Fig. 1. The first stage is known as the primary or input transducer, following are the secondary or intermediate stages and, finally, there is an output transducer.

In the fuel tank a float transduces the fuel level to an equivalent rotary motion. This drives a rotary potentiometer which provides a voltage proportional to the angle of rotation. Sometimes there is a calibration or adjustment stage in the chain. At the dashboard the voltage is turned back to a rotary displacement in the fuel gauge meter movement. The advantage of the electrical signal is that it avoids the need for a complicated mechanical linkage between the fuel level and the gauge. In a control application an electrical measurement output signal also enables in-line correction, compensation and computation to be made before the signal is used. Recording is also made most easily with electrical plotters.

In principle, a transducer is a simple device. In practice, however, simple schemes invariably suffer from defects that limit the ability of the device to provide repeatable and accurate values.

They may suffer from wear as time proceeds: environmental factors such as temperature, pressure, humidity and shock, for instance, may be a significant problem. Consequently, at first sight, the developed transducer system usually appears quite complicated. But if treated systematically, it can be broken up into separate sub-systems that perform distinctly different tasks, each being joined to produce a satisfactorily reliable and accurate device.

A list of all the different transducers yet devised would be never ending, for the basic physical effects that could be used are beyond complete classification. Each may be used for many different purposes. For example, a light spot moving across a photo-cell can be used to measure position, alternatively, the movement might be used to change the sound level of a radio receiver by varying the voltage applied to the receiver output stage.

Nevertheless some transducers have emerged that are well developed for specific tasks. Thus a brief list can be made of primary devices, and those quantities measurable by the use of intermediate stages.

Linear Movement: From this are also derived thickness, velocity, acceleration, force, wear, vibration, hardness, stress, strain, pressure, gravity, magnetic field, level and position, by the use of secondary devices.

Angular Movement: Angular vibration, tilt, torque, and position are obtained with angular transducers.

Temperature: Flow, turbulence, heat conductivity, remote sensing and displacement can be obtained by use of this basic measurement.

Illumination: Length, force, strain, torque, frequency, and light distribution have been measured using illumination.

Time: Speed, counting, frequency and position rely on time measurement.

Force: Weight, density, stress, torque and viscosity use force indirectly.

It is many years since James Watt thought of using the speed indicator of an early steam engine automatically to control its speed, and so producing what was probably the world's first industrial feedback control system to find extensive use.

In that case, a centrifugal governor was used to change the difficult-to-detect shaft speed into an equivalent mechanical displacement. It was, in fact, what is now called a transducer.

Since the time of the industrial revolution, machines and processes have developed at an ever quickening rate and the need to convert difficult-to-use effects into alternative physical forms has grown rapidly.

Late in the 19th. century, electricity became available to industry and science. Then the electronic discipline emerged. Electronic techniques, allied with those of mechanical, optical, thermal and acoustic origin — the list is never-ending — enabled a vast array of transducers to be developed to fulfil the needs of sophisticated measurement and control.

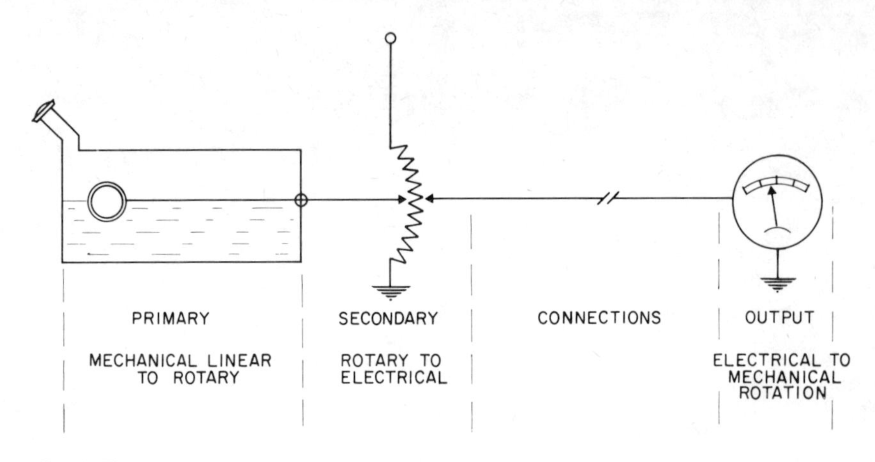

Fig. 1. The stages of transduction in a fuel-level gauge.

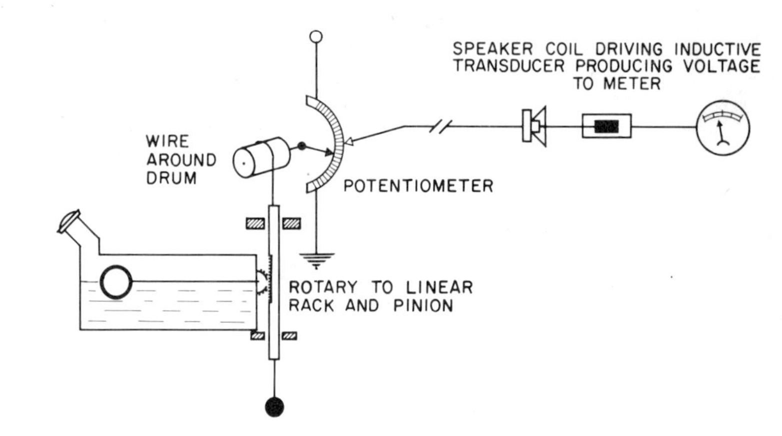

Fig. 2. A circuitous solution to a problem.

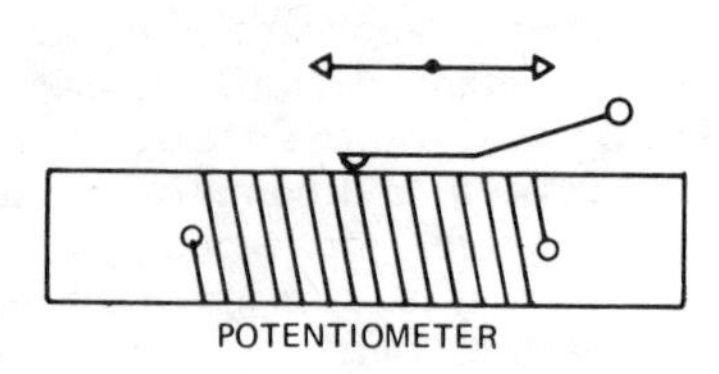

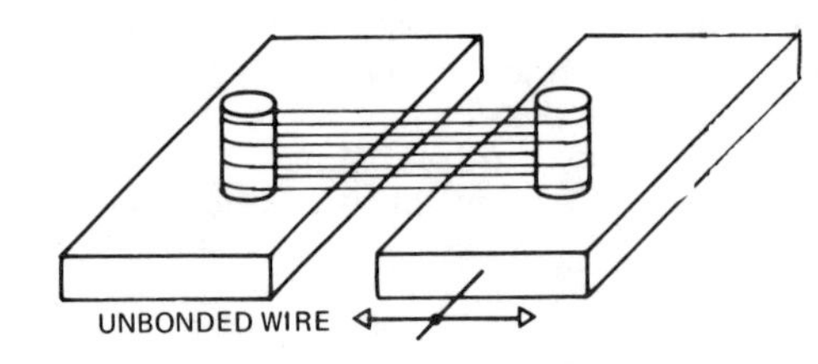

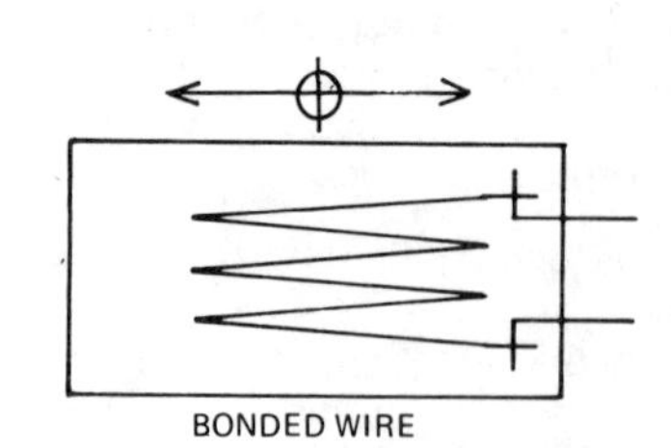

Fig. 3. Resistive displacement principles

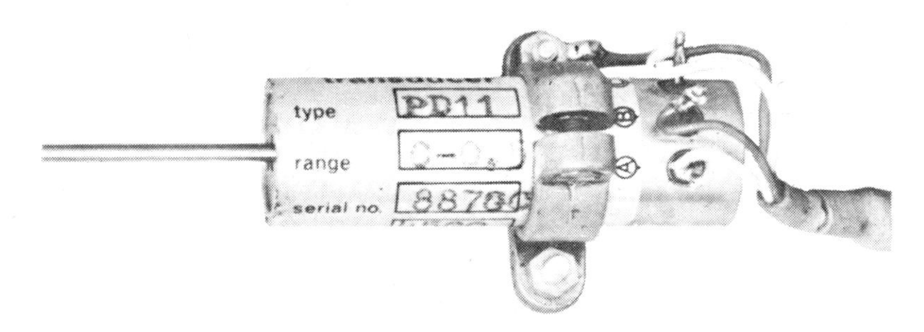

Fig. 4. A linear wire wound displacement transducer having 25μm resolution.

This list is not complete but it does illustrate the variety of possibilities open to the designer. A problem can be solved by circuitous means (Figure 2 is a fuel gauge arrangement with redundant use of transducers) but economic and reliability factors decide which way is acceptable in reality.

Transducers may provide the transduction in one of two basic ways. It may, firstly, control the available source of energy as a tap lets water through or a variable resistance controls the current flow from the power source in a circuit. Secondly, the transducer may actually convert the original energy form into another more appropriate form. An example is the use of a photo-voltaic cell in which light radiation energy generates electrical energy. Transducers may also provide mechanical energy from the available electrical source as happens in the moving coil loudspeaker.

An interesting fact is that the dynamic and static behaviour of mechanical, acoustic and electrical systems are each described by similar mathematical equations. This analogy, as it is called, enables the behaviour of large machines to be simulated by inexpensive electrical networks. For example, the internal-combustion engine can be simply represented by a resistor and a capacitor at speeds above idling. So for research purposes, once the value of R and C are determined, it is possible to study the performance of that engine in a computer.

Some transducer applications need only a slow speed static response but often the need is for rapid conversion. The frequency response is, therefore, of interest. Mechanical systems are generally incapable of the same high speeds obtainable in electrical devices. For this reason there is a trend toward total electronic techniques if possible. This is not always a prudent way to solve the problem as many mechanical devices have been extensively developed to provide reliabilities of years (or millions of operations). A simple example is the choice made when several independent circuits have to be switched together. A bank of reed-relays is inexpensive, simple to design and capable of excessive overloads. A solid-state equivalent circuit may be more expensive to develop and more prone to overloads. Each case should be considered on its merits.

Several terms, commonly used in measurement are often misunderstood and misused. The first is the **repeatability** of measurement. If repeated measurements are made of a static process by an instrument with sufficient sensitivity there will be a **scatter** of the values around some mean value. This scatter represents the uncertainty of the measuring process used. The most commonly used method of expressing this scatter is by what is known as the standard deviation (σ). This is found by a simple statistical mathematical formula. The important thing to realise is that there is a 68% chance of the true value lying between plus and minus 1 σ. For example, if a voltage is measured 100 times and its mean value found to be 100V with a standard deviation of 2 volts this means that 68 times it will lie between 98 and 102 volts and 32 times it will be outside these limits. In practice, one-σ limits are not tight enough. For $\pm 2\sigma$ limits it is 95 times out of 100 and for $\pm 3\sigma$ limits 99.7 out of 100 times within. Repeatability is the first requirement of a transducer for without it accuracy has no meaning. (The standard deviation of any transducer or precision measuring instrument is almost always quoted by the manufacturer.)

The **resolution** of a measuring instrument is the smallest quantity that it can detect. But to have extreme resolution does not imply that it will repeat each time nor be accurate. A screw-thread micrometer could have a drum of enormous diameter enabling extremely small distances to be gauged, but the screw friction and error would produce scatter and inaccuracy.

Precision is the term used to describe how well the instrument measures and gives a reliable value. The smallness of the standard deviation, therefore, is a measure of precision.

Accuracy is the most difficult factor to obtain. An instrument may be precise, always giving the same value, but to be accurate, that value must be true to the established standards. For example, a voltmeter may indicate

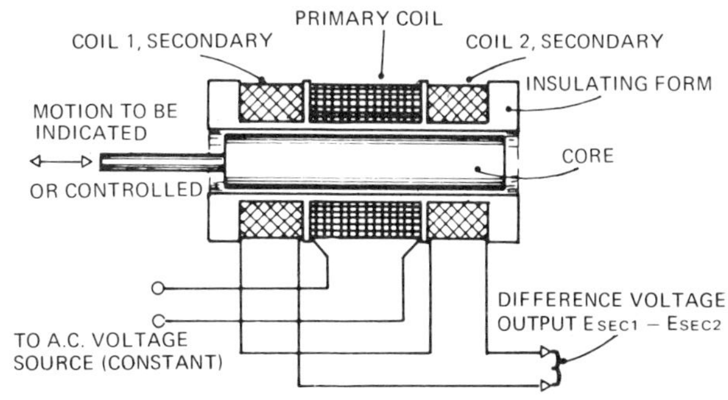

Fig. 5a. Linear variable differential transformer.

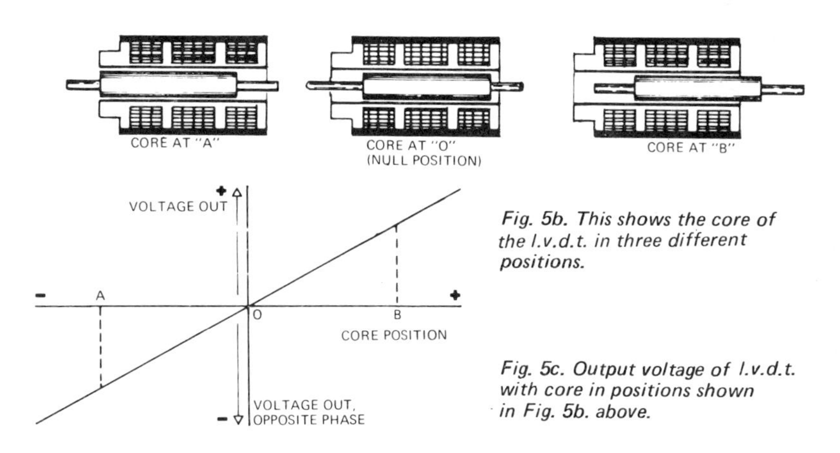

Fig. 5b. This shows the core of the l.v.d.t. in three different positions.

Fig. 5c. Output voltage of l.v.d.t. with core in positions shown in Fig. 5b. above.

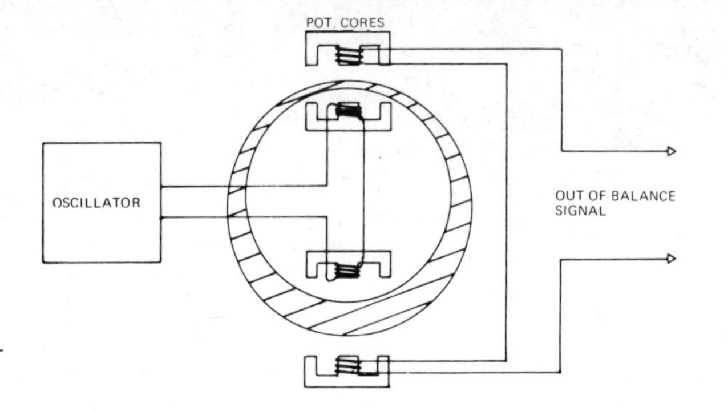

Fig. 6. A differential arrangement using two variable reluctance transducers for monitoring tube eccentricity.

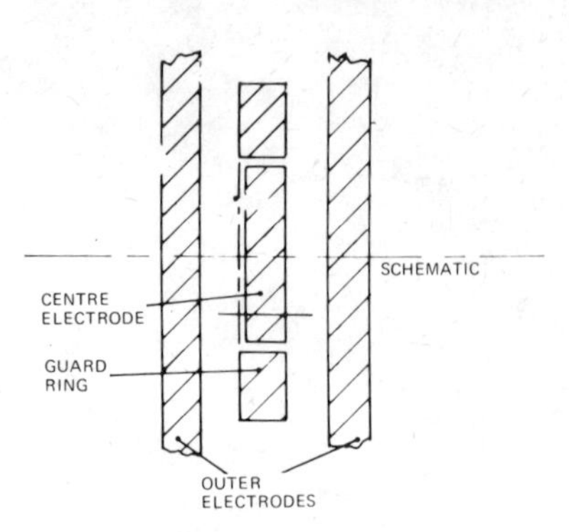

10.1 volts repeatedly but if the pointer is bent or the multiplier resistor incorrect, the actual voltage may be only 9.5. There is no way of establishing accuracy without referring to another measurement device. Often, accuracy is added to a precision instrument by resorting to calibration. In transducer application, this must usually be automatic, or built into the device, as a human link is undesirable.

So much for a general basis of transducer technology. We will now continue by discussing various measurements in turn starting with the methods used to transduce displacements.

This initial chapter deals with small displacement transducers ranging in capability from a few millimetres down to hundredths of the diameter of atoms. These devices are particularly useful in obtaining derived quantities as well as direct measurements (as will be seen later). In the second chapter we shall discuss the industrial displacement range, that is, from millimetres to several metres, and then the surveying range from hundreds of metres to the size of the Earth and larger.

MICRODISPLACEMENT TRANSDUCERS

Displacement is measured directly with resistive, inductive and capacitive methods and, indirectly, by optical means.

Resistive: The simplest way to transduce movement into electrical signals is mechanically to vary the properties of a resistance. This can be realised by direct mechanical movement of the tapping point, as in a potentiometer, or by straining the resistance element, as in a strain gauge, (Fig. 3).

Potentiometers, whether linear or rotary, consist of a resistance track upon which slides a contact wiper. The earliest precision potentiometers used fine resistance-wire wound around a toroidal former. As the wiper moved over the turns, the output changed abruptly and this limited the resolution. A modern type linear potentiometer is shown in Fig. 4. Continuous resolution has been obtained by using a slider running longitudinally along the wire (it may also be obtained by the use of composite-material track). Repeatability is limited by the precision of the wiper contact position and slight variations in electrical contact. Due to relatively poor repeatability and reliability, and because of the high operating force, it is unusual for a resistance potentiometer of this type to be used for applications where high resolution is required.

An unusual type of resistance potentiometer is that whereby a tightly coiled tension spring is stretched to open the coils and increase the resistance. The required displacement-output signal characteristic is determined by the method of hard-coiling the spring.

The sensitivity of resistive methods is limited by the allowable self-heating of the element, for temperature changes alter the resistance value.

Strain gauges are resistances that are strained bodily so as to alter their physical cross-section and length. Resistive types are made as wire, or stamped foils of thickness around 20 μm, and are arranged to obtain multiple elongations connected in series. Adhesives are used to attach the gauge to the member to be measured. This ensures faithful movement with the parent. Typical resistance values range from 10 to 10 000 ohms. Self heating and temperature effects limit the sensitivity of these devices but absence of mechanical moving contacts enables resolutions of better than a microstrain to be obtained.

The ratio of strain to proportionate resistance change is termed the gauge factor. This is usually quoted by the manufacturer. For linear resistance gauges it is close to 2.0. Calibration is necessary for precision work.

Wheatstone bridges, of simple and advanced form, are used to measure the resistance changes of both potentiometers and strain gauges. To compensate for temperature, a dummy resistance is used in one arm of the bridge.

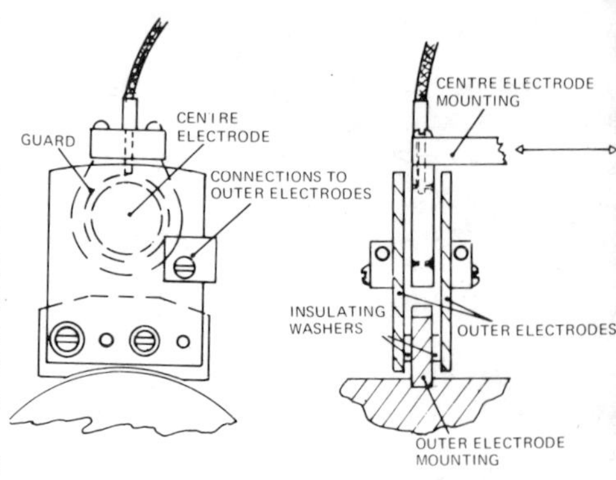

Fig. 7 Differential parallel plate capacitance transducer developed at the National Measurement Laboratory, Sydney.

The main advantage of resistive strain gauges is their extremely small size, ranging from 2 mm upwards. Frequency response exceeds 50 MHz for special, surface deposited types.

Solid-state strain gauges are also available. If a semiconductor element such as silicon is strained, it also shows a change in resistance. Their gauge factor is not constant but depends upon instantaneous strain magnitude and temperature. Gauge factors of 100 are typical.

The main disadvantage of resistive strain gauges is their fragility, and this requires them to be mounted on a more substantial element. For fixed applications, it is practicable to mount the gauge between the two moving members in what is known as an unbonded arrangement.

Strain gauges are used extensively in civil and mechanical engineering testing. Gauges are glued to the structure in many places. A data-logger reads each in turn recording the strain at that time. These data are then processed to produce the required information.

Inductive: Electromagnetic and electrostatic fields can be utilised for displacement sensing, each having practical advantages. Alternating current excitation can be employed and dissipative circuit elements are kept to a minimum (factors which enhance sensitivity and reduce drift). Inductive methods use, in the main, either the linear variable differential transformer (l.v.d.t.) principle, or operate on a reluctance variation concept.

The l.v.d.t. consists of a spatially centre-tapped solenoid in which moves a magnetically-hard steel core, (Fig. 5). The coil is energised either by a separate primary coil or by direct connection across the winding. As the core moves relative to the winding the flux-linkages cutting each half of the winding vary, resulting in amplitude unbalance between the halves. The degree of unbalance is linearly related to the core's displacement from the coil centre.

One method of sensing the unbalance is to connect the sensing coils in opposition and measure the output voltage. It is necessary, however, in this simple method, to determine the phase relationship between the excitation and output in order to decide the sign of the displacement. A superior technique uses a phase-sensitive detector, the output then being a bi-polar dc voltage which is linear with displacement.

Linear variable differential transformers are used extensively in industry in weighing machines, pressure transducers and load cells; and in science in earth strain-meters, tilt meters and seismometers. A major manufacturer offers over 2000 different models. In these applications resolution required is rarely less than 5 μm.

The principle is also used in some industrial dimensional metrology gauging heads where 100nm is the best resolution needed.

The core and winding are mounted to avoid mechanical contact, but perpendicular movement to the core's axis is not possible. Axial core travel can be over very large distances and the zero position can be set electronically at any point along the length of the winding. Humidity, even liquids, do not affect the operation. Magnetic shielding is used to isolate the winding from external fields.

The other main inductance technique employed is known as the reluctance transducer. If the air-gap of a magnetic circuit is varied, the magnetic circuit reluctance changes. As the majority of the circuit reluctance is produced across the air-gap the response is reasonably linear. In practice the iron circuit can be made from a pot-core as shown in the tube gauge (Fig. 6). This contains the sensing coil and a freely

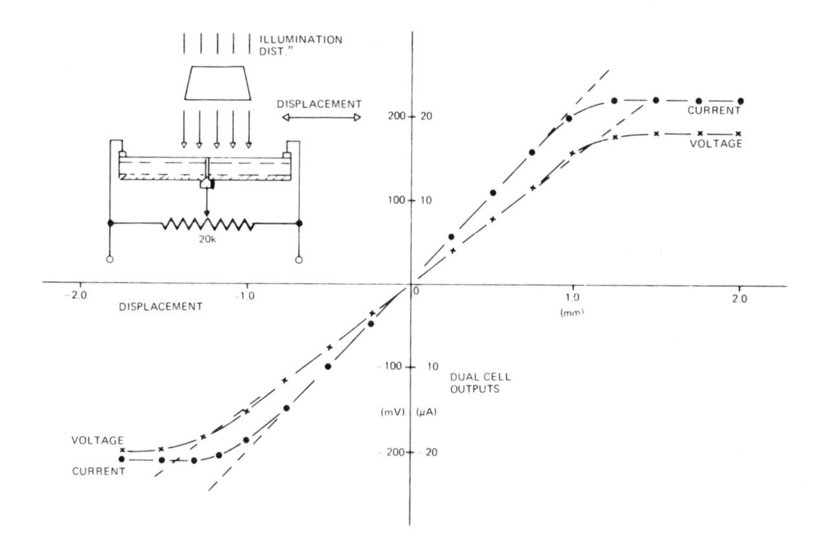

Fig. 8. Dual cell position-sensitive optical transducer.

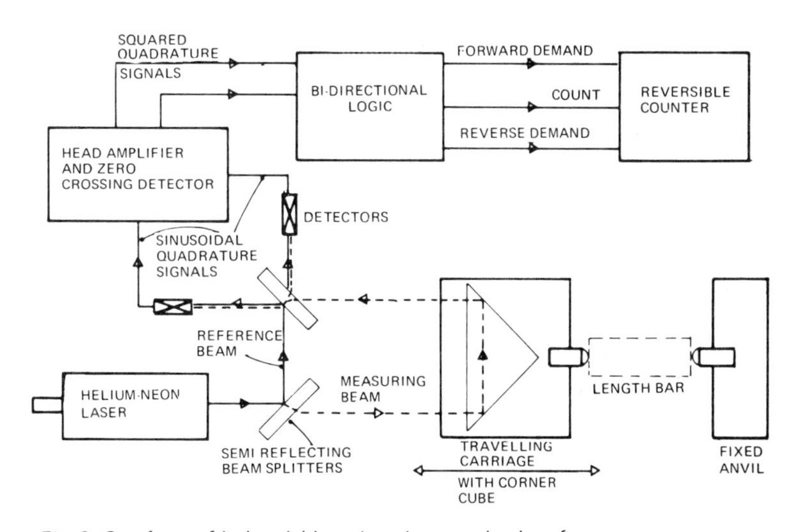

Fig. 9. One form of industrial laser length measuring interferometer.

moving limb which completes the circuit. The device is directly energised and may be sensed by similar methods to l.v.d.t.'s. A differential arrangement is often used to balance the effects of temperature and stray fields.

Reluctance transducers have been employed for measuring tube eccentricity as shown, for measuring dynamic lubricant film thicknesses, and in pressure gauges. Research on a borehole tilt meter at the Australian National University used a reluctance technique to sense the pendulum movements. Sensitivities of these small range inductive methods can be as high as 200 mV/μm. Frequency response is limited by the excitation frequency used (10 Hz - 10 kHz) and mechanical factors.

Capacitative: The most favoured extreme precision sensing method is known as capacitance micrometry. In its simplest form it consists of measuring the capacitance changes resulting as the separation between two plates of a capacitor is varied. As capacitance is inversely proportional to gap distance, the displacement/output characteristic is a non-linear hyperbolic. A guardring is used to control fringing of the electrostatic field existing between the plates and to reduce the effect of lead strays which shunt the small variable-capacitance limiting the attainable sensitivity. Linearization has been achieved in one manufactured gauge by placing the sensing capacitance in the feedback of an operational amplifier.

The magnitude of the sensing capacitance is only a few picofarads. Reactive bridges can sense to 10^{-5}pF, or a little better, using tap changing inductive transformers. As the capacitance value is proportional to plate area and inversely to separation,

Fig. 10. Testing the corrosion thickness of a pipe with an ultrasonic gauge.

highest sensitivities result for largest plate sizes and smallest gaps.

Practical considerations of plate flatness and degree of parallelism limit the gap size to around 100 μm or more. Plate diameters in use range from millimetres to centimetres. In most applications the sensitivity of the method to stray capacitance is reduced by using a differential capacitance mechanical layout. A central plate moves between two fixed sensing electrodes, the plate being earthed. Any temperature effects and air dielectric changes occur equally in each arm of the arrangement. If sensed by a bridge circuit, these effects are largely cancelled. A unit developed at the National Measurement Laboratory in Sydney is shown in Figure 7.

Capacitance gauges have been used in geophysical instruments such as gravimeters, tilt meters and strain meters. They are also used in industrial gauging and machine tool control.

Optical: Mechanical displacements of interest can be converted into movements of a light beam which can then be sensed with a position-sensitive optical detector. Rotations can be magnified using an optical-lever if space permits.

In simple arrangements the radiation beam is either split into halves, each half feeding a separate photocell or alternatively, the beam may impinge directly onto a photo-device with position-sensitive characteristics. In each case a differential bridge arrangement is usually incorporated giving zero output if the beam is truly centred. This null position can be conveniently displaced by electrical means.

In brief, static arrangements use position-sensitive photo cells or passive optico-mechanical arrangements (beam splitting mirrors, prisms) and dynamic methods use optico-mechanical devices (rotating prisms and wedges, vibrating apertures) or electrodynamic devices (image dissector tubes, magneto-optical and wavefront shearing).

Numerous possibilities exist, but for simplicity and cheapness, solid-state position-sensitive photo-cells will usually be the first choice considered. The simplest method uses two (or four for 2 axis measurement) silicon solar cells, about 10 mm square, which are mounted adjacent to each other. This is illustrated in Figure 8. A rectangular light spot is traversed across the junction. If central, each produces an equal signal which cancel if they are differentially-connected; this is the null position. Displacement from the null gives a proportional output until the spot moves entirely onto a single cell where a saturated displacement characteristic occurs.

In 1957 a lateral-effect position-sensitive photocell was reported in which the output is logarithmically related to the spot displacement as it moves between two ohmic contacts made on the junction surface. Extensive research was concentrated on these cells for tracking of military targets such as the plume of a missile.

A third form of position sensitive cell uses the light-spot as a contact 'wiper'. It effectively shorts a low-impedance track, via a photo conductive strip, to a position along a high-impedance potentiometer track.

In most of these optical position-sensing methods it is paramount that the beam intensity remains constant as output away from the null (at the null point intensity is less important) is proportional to the luminous flux falling on the cells. This in turn, is decided by the total beam flux and its distribution.

Another way to detect position is to have an array of photo-diodes, interrogating them to find the position of a spot or a pattern illuminating them. Arrays containing 2500 diodes have been made.

These optical methods can detect movements perpendicular to the beam's axis. Interferometry can detect movements along the axis to extreme precision.

If a coherent radiation source is split into two paths, each being optically mixed upon return from reflectors, the position of the interference fringes resulting is a direct measure of length differences between the two arms. If one arm is fixed as a reference length, displacements in the other arm can be measured by monitoring the fringe movements. A unit developed in Britain is shown in Figure 9. Suitable radiation wavelengths range from millimetres to micrometres, so in most cases the monitoring task involves whole fringe counting and then fringe width subdivision or interpolation. The shortest practical wavelength is around 500 nm, which in the simplest interferometer accounts for 250 nm displacement of the measuring arm.

One well-used method of interpolation is to produce two signals from the fringes which are 90° spatially separated. Digital operation on these dc coupled signals will give four pulses per fringe. This technique was developed simultaneously in 1953 for interpolation in an interferometer and in Moire grating use for industrial control by Ferranti. A number of totally electric methods have been devised to obtain improved resolution from dc quadrature-phase signals. These include mechanically activated sine and cosine potentiometers driven to balance, use of resistance networks to produce a set of different phase triangular signals which can be divided by trigger levels and super-position of the signals on to an ac carrier which then enable phase-sensitive detection to be used. At the best, however, these methods are normally only accurate to within 1%.

Another way to interpolate the fringes is to drive the return mirror so as to maintain the fringe in a constant position. This method has been used in the University of Cambridge laser earth strain meter. In all cases of fringe monitoring, however, it is possible for optical and electronic noise to displace the fringe too rapidly for the system to record, thus losing or gaining an integral number of error counts.

Laser interferometers are used in industry for the exacting calibration of jig boring mills and the like. With the industrial units, the effects of the air (that is the change in temperature, humidity and pressure) limit the precision to around 1 part in a million. This is improved by feeding back data on the conditions using appropriate transducers. In some applications, notably earth strain interferometers, the complete system is contained in an evacuated tube to avoid these errors. In such cases, precisions of around 1 part in 10 000 million are realised if the wavelength of the laser is stabilised.

Miscellaneous:

The above are the most popular methods for sensing small displacements. There are many other ways to solve the problem and each has its particular attributes which make them suited to special applications. Here are just a few.

Radiation Gauging — Here a source of short wavelength radiation (α, β or γ) is located on one side of the (thin) material to be measured. The degree of absorption, measured by a radiation counting detector on the other side, is a measure of thickness. A number of variations exist on this, for example, shuttered absorbers are used to measure axial displacement in turbines and one-side gauges have application in continuous thin plastic-film measurement. The measurement precision depends upon radiation count integration so accuracy is increased by averaging the count over a longer period.

Ultrasonic Gauging — If the velocity of propagation is known, the transit time of an acoustic wave within a material is a measure of thickness. Sound waves travel at about 300 m/sec in air, 1500 m/sec in water and 5000 m/sec in metals. This principle has been used for small distance gauging. The slower velocity of electro-magnetic radiation, enables finer resolution to be obtained for a given technological limit on transit time measurement. An extensive study of an ultrasonic micrometer has been made at the Atomic Energy Research Establishment in Britain where they have developed units that resolve to 2 μm. Ultrasonics have been successfully employed for engineering component thickness measurement, corrosion thickness measurement in pipes, (see Fig. 10), and for medical applications in which foreign objects are located, growths discovered and probes guided.

Laser Beam Diffraction — Coherent radiation diffracts around a small object to produce an interference pattern beyond it. This has been used to gauge wire size diameters down to 10 μm. The position of the chosen diffraction fringe, (best produced by a laser source) can be monitored by a position sensitive photocell to enhance the resolution. This method is capable of size measurement at very high speed.

Sub-millimetre Waves — In many applications of interferometry the wavelength of the source is too short compared with the surface finish to be gauged against, and a mirror must be added. The National Physical Laboratory in England have developed an interferometer using submillimetre waves of wavelength 50-1000 μm. Their device has been called the Teramet. It can measure to normal tight engineering tolerances (2 μm) but needs no specially-provided reflector as in laser interferometry.

Other lesser known techniques include vibrating-wire strain gauges in which the tension of a continuously vibrated wire is varied. (The resonance frequency is then a measure of length change causing the tension change); piezo-electric crystals in which a force (accompanied by very small proportionate compression or extension) produces an electric charge flow which can be calibrated as displacement; pressure sensitive paints and semiconductors that exhibit resistance changes as they are deformed mechanically; and the use of a television pick up tube (usually the vidicon) to produce serial electrical signals of an optical shape enabling amplification to be achieved and an electrical output to be obtained.

FURTHER READING

(see also lists of chapters 2-5)

"Handbook of Industrial Metrology". ASTM, Prentice-Hall, Englewood Cliffs, N.J., 1967.

"Engineering Measurements". B. A. Barry, Wiley, New York, 1964.

"Mechanical Measurements". T. G. Beckwith and N. L. Buck, Addison-Wesley, Reading, Mass., 1969.

"Dimensional Metrology". I. H. Fullmer, NBS, Washington, 1966.

"Engineering Metrology". K. J. Hume, MacDonald, London, 1970.

"Engineering Dimensional Metrology". L. Miller, Arnold, London, 1962.

"Foundations of Mechanical Accuracy". W. R. Moore, MIT Press, Cambridge, Mass., 1971.

"The Strain-Gage Primer". C. C. Perry and H. R. Lissner, McGraw-Hill, New York, 1962.

"Metrology and Precision Engineering". A. J. Scarr, McGraw-Hill, New York, 1968.

"Engineering Metrology". G. G. Thomas, Butterworths, London, 1974.

CHAPTER 2

INDUSTRIAL AND SURVEYING RANGE LENGTH TRANSDUCERS

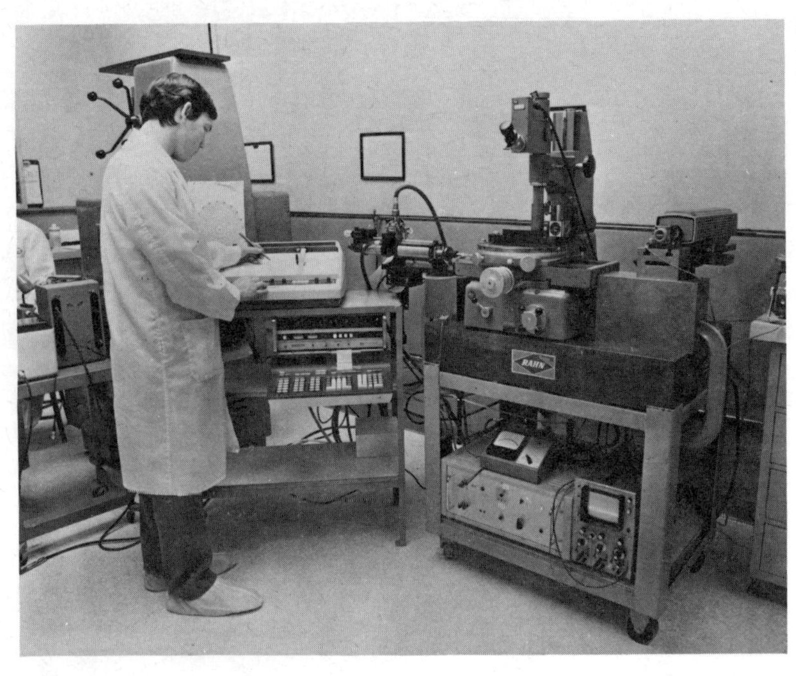

Dimensional-measuring interferometer systems have provided wide-range, extreme precision capability. (Hewlett-Packard)

THE first chapter dealt with methods for converting small displacements into electrical signals. This time, longer length transducers are discussed. With only a few exceptions, the previously considered methods are unsuited to ranges greater than millimetres, so other ways have been devised.

The majority of precision length measurement of distances ranging from millimetres to several metres is performed in industry, so we use the term industrial range to assist classification.

Distances greater than 100 m or so are grouped in what could be called the surveying range, as it is mainly for land survey purposes that long distance measuring instruments have been developed.

INDUSTRIAL RANGE DISPLACEMENT TRANSDUCERS

Prior to 1950 electrical length-transducers were not often used in general industrial practice. Instead measurements were made with manually operated instruments, many having been devised to cope with specific measurement tasks. Examples are gear testing machines, projection microscopes, travelling microscopes and gauge interferometers.

Then came the change. Groups in the United States of America and in Britain foresaw the potential of an automatic machine tool that could produce a variety of different components at the command of taped digital signals. Numerical control (N.C. for short), was the subsequent development that has been accepted throughout the world.

One of the larger numerically controlled mills is shown in Figure 1.

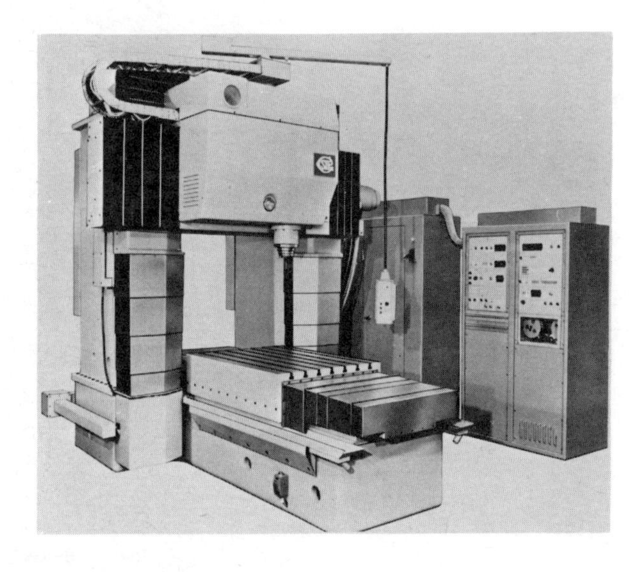

Fig. 1. High-precision numerically controlled machine tool. (SIP)

Control techniques were reasonably well understood due to 1940's development of gun positioning and radar tracking, but at that time no transducer had been developed that could provide an electrical indication of a machine tool's slide position. Such a transducer needed to have a range of around a metre and a precision and accuracy close to a few parts in a million.

It did not take long for the necessary technique to be developed, for by 1955 there were dozens of such transducers in existence. Other uses for these transducers were exploited as their benefits were realised.

The simplest purpose for which they can be used is to assist the machine operator by providing a readout of length. As most devices provided a digital form of readout rather than an analogue indication, the term digital readout, or D.R.O., came into use. (One commercial D.R.O. unit does, in fact, display with a rotating meter). A milling machine with a D.R.O. facility is illustrated in Figure 2.

The trend of D.R.O. and N.C. spread to other applications. Draughting machines, chart digitizers, oxy-acetylene plate cutters, frame benders, tube benders, wiring loom production machines, rolled steel joist drilling machines, locomotive door welders and even cranes in a steel yard have had transducers fitted for readout or control purposes.

It is not ideal, however, to fit the transducer on the way of a machine, for machine constructional inaccuracies exist between the slide and the cutting point of the tool.

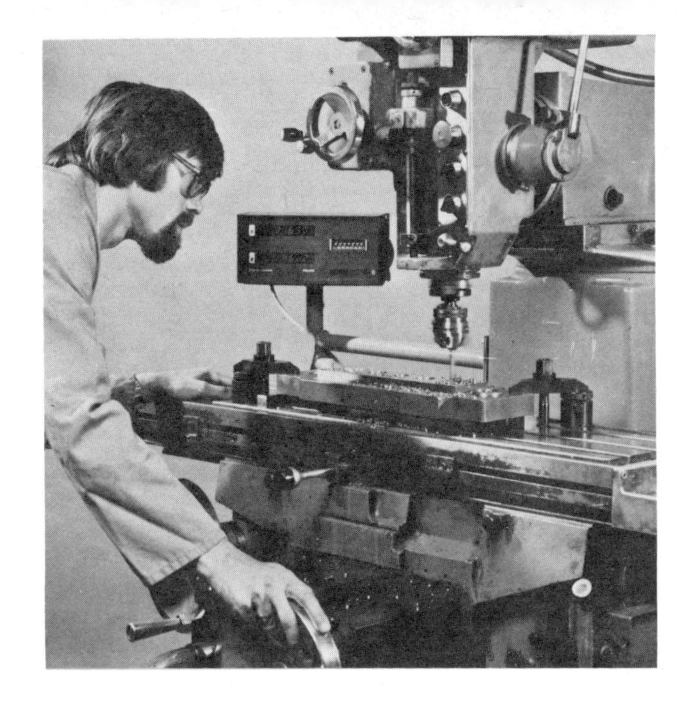

Fig. 2. This vertical milling machine has a digital readout system to assist the operator by displaying the position of both traverses.

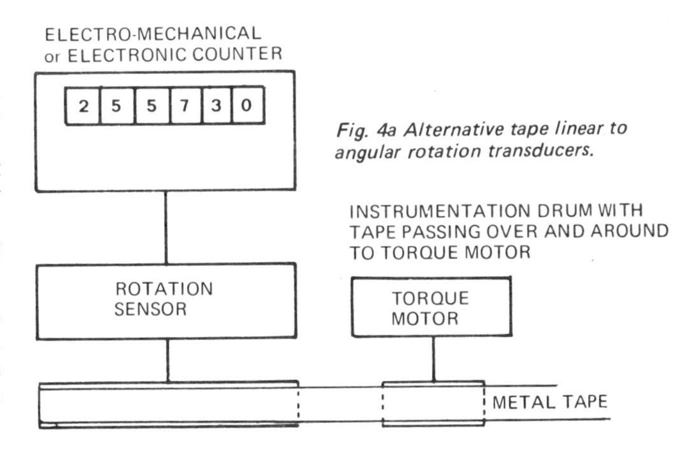

Fig. 4a Alternative tape linear to angular rotation transducers.

Fig. 3. Rack and pinion installation on an Innocenti boring mill.

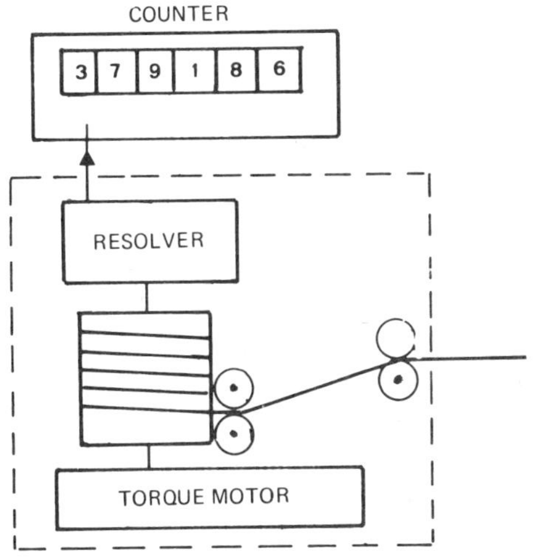

Fig. 4b Alternative wire linear to to angular rotation transducers.

An axiom of measurement is Abbe's principle (after Professor Ernst Abbe of the famous Zeiss optical works, who lived in the 19th century). This states that a dimensional measurement made directly between the points of interest is better than one made by transferring with linkages or the like. This is illustrated by considering the measurement of length with a ruler. If the ruler is placed right on the marks, no parallax error is produced as no perpendicular transfer is needed. If the ruler is placed in line with the marks and moved from one end to the other, transferring errors are also avoided. Although this is an obvious principle, it is often impracticable to observe it as measurement is but one of the functions to be considered when designing a machine-tool.

If the size of a part is measured right at the work face, it is possible to eliminate machine structural errors. This idea has been termed the in-process technique and has found application in lathe work where a type of micrometer measures right at the tool-bit as work progresses.

Although the majority of current applications for industrial range transducers are in industry, they are not restricted to the workshop alone. In the 50's the developments in N.C. prompted many people to claim automation was around the corner. Today there are fully automatic manufacturing systems, especially in automobile production, but in the main these are not automatically controlled, but rather are preset mechanically to produce the same part many times over. Machining centres, as the fully automated systems are called, have been made and technologically, automation is possible. Social and economic pressures have prevented their greater use so far.

There are literally thousands of N.C. tools and D.R.O. units in use throughout the world so it is only to be expected that numerous ways have been devised to transduce length into control signals. Two basic approaches to the problem are possible.

In the first, the linear motion is converted into a rotary equivalent by a mechanical method. This rotation is then transduced to give either an analogue or digital measurement signal.

The alternative method utilises measurements taken from a directly sensed linear scale attached — where mechanically convenient — to the machine.

LINEAR TO ROTARY CONVERSION

There are four mechanical devices that can convert length to angle over long distances. These are the rack and pinion, lead screw and nut, tape or wire and drum, and a friction driven wheel running on the linear surface.

Rack and Pinion — A popular technique, especially for long traverses on machine tools, uses a precision gear pinion meshing in a linear gear track which is mounted on the slide (as shown in Figure 3). Provided the mesh is accurate and back lash controlled, this method can provide accuracies around 10 parts in a million (which is the generally needed workshop accuracy). The design of the pinion gear and the pinion mounting is important, and usually springloaded split gears are used to minimize backlash.

Leadscrews — Early screws left much to be desired as backlash and friction were considerable. Nowadays the friction screw has been replaced in precision designs by the recirculating-ball screw in which ball bearings maintain contact between the screw and the nut. The nut is made in two pieces, one being wound on against the other to preload the balls into heavy contact in order to increase the stiffness of the joint. This results in improved dynamic performance. Ball screws are expensive but yield excellent precision. Better grades hold tolerances of 2μm/250 mm of screw. The length of screw is limited, however, to a metre or two by the difficulties of supporting the screw and by the amount of screw wind-up under load that can be tolerated.

Tape and Wire Driven Drums — This linear to rotary conversion makes use of a tape or wire to rotate a precision measuring drum. The tape may be pulled around the drum, using the drum as a capstan (Fig. 4a), or pulled off the drum. The latter method uses a spring or an electric motor to provide a constant torque to the storage/measurement drum (Fig. 4b). This maintains a constant tension in the wire or tape and reduces elastic errors.

Although this method can provide accurate measurements, comparable with the rack and pinion for instance, it is not widely used except for the measurement of fluid levels in storage tanks. An automatic positioning

Fig. 5. Friction driven diameter measurement of a large shaft in a lathe (Rotax Ltd).

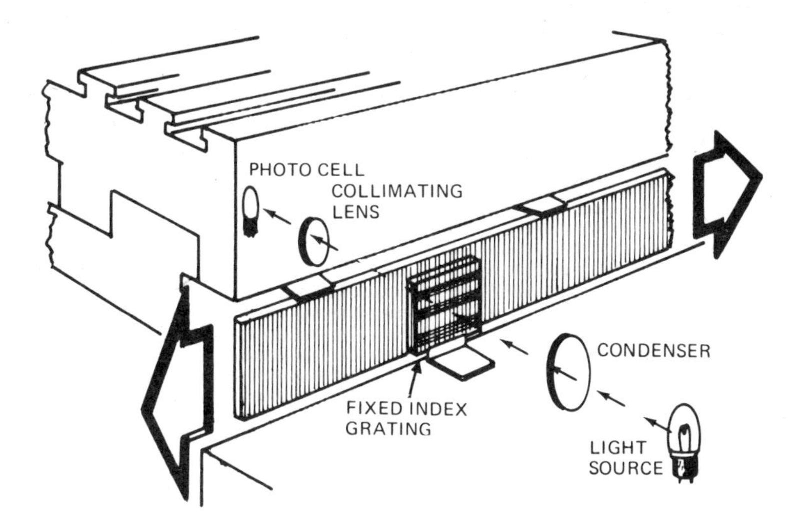

If two diffraction gratings consisting of ruled lines on transparent blanks are superimposed with their rulings not quite parallel to each other, dark bands called Moire fringes are produced across the gratings. The spacing of the fringes is a multiple of the ruling interval and is dependent on the angle of inclination between the two sets of lines.

Movement of one grating relative to the other, in a direction perpendicular to the lines, causes the fringes to move in a direction at right angles. If, for example, the grating is moved a distance equal to the ruling interval the fringes move a distance equal to one fringe spacing, and so provide an amplified indication of the movement of the grating. In this effect lies the principle of the application of gratings to the requirements of engineering metrology.

Fig. 6. Simplified arrangement using bar-space optical transmission gratings to measure displacement.

control system has been demonstrated that uses one of these units. As only two ends (the end of the wire and the tensioning drum) need to be mounted alignment is far less critical than with alternative methods.

Friction Rollers — It is possible to convert a linear motion into a rotary one by using a wheel. Provided the surface is smooth and flat this method yields quite good precision. An inexpensive clock-dial output device is available for fitting to a lathe. It is also available with a digital readout device.

Friction driven rollers have also been used to control shaft diameter size in in-process turning and grinding. To improve the accuracy, rotations of the roller (mounted in contact with the surface just cut) are counted and subdivided over a large number of shaft turns. Devices developed in Britain perform better than a micrometer and, of course, have the advantage of an electrical output for inspection or control purposes. The advantage of this method is that the same basic unit can measure from small to practically unlimited sizes of shaft (see Fig. 5).

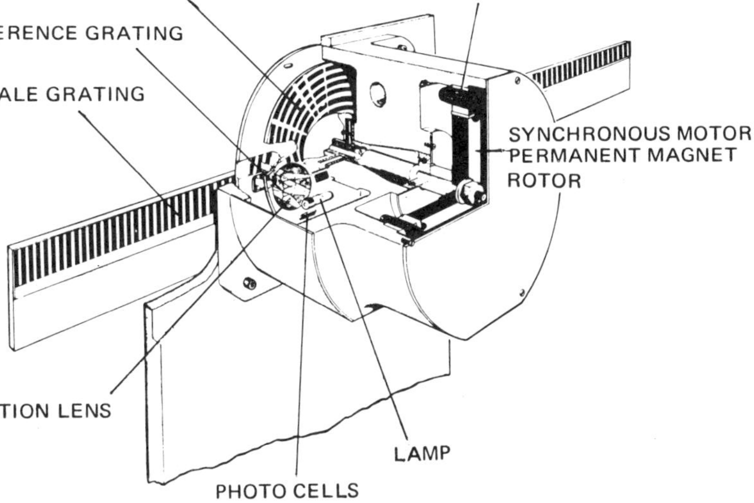

Fig. 7. Mechanical modulation of optical grating signals in the Ferranti spiral scanning head.

In each of the above methods the resultant shaft rotation is used to rotate a mechanical pointer or an angle transducer. Angle transducers are discussed in the next chapter.

Direct Methods

There is often a requirement for greater precision than indirect methods can provide.

Accuracies approaching a part in a million are attainable with directly sensed linear scales. Extreme precision is, however, often demanded by persons unaware that errors of one part in a million are hard to eliminate unless the whole machine and work-piece are temperature stable to at least 1°C or better. Few machines, even precision ones, are given a special controlled environment room in which to operate.

Scales can be either a physical mechanical arrangement or a feature of a radiation beam such as in a laser interferometer.

Mechanical scales can be sensed by the primary electrical methods (resistive, inductive or capacitive) or by optical methods. In some instances the scale is simply a length with subdivision marks made along it in some way. The marks may be individually identifiable or they may all appear identical. These alternatives are known as absolute and incremental scales respectively, and the sensing technique depends much upon which type of scale it is. In the incremental version, the marks are progressively counted in order to ascertain the length traversed, whereas in the absolute scale, no action needs to be taken during movement between two points as the position information is available at the mark on which it stops.

Resistance potentiometers, described in the last chapter, can be made to any length but for applications requiring extreme precision, the resolution and stability are inadequate. Furthermore industrial applications offer extremely dirty and vibratory conditions which severely shorten the life of the contact and surface of the resistance material.

The inductive and capacitive techniques described in the last chapter are only suitable for quite small displacements. However, means have been developed by which a number of units can be cascaded side by side to cover the required length. Electronic circuitry is used to decide which unit is in operation at any one time, thereby giving the coarse position. Fine position is added to the measurement by using the output of the individual transducer then in use. This is a combination of both incremental (the coarse positions are identical) and absolute devices – it is known as a hybrid system.

Many variations exist on this theme. In one, which illustrates the general principle used, a photo-mechanically printed and etched conducting hairpin winding, is formed on the surface of a precision glass plate some 300 mm long. This is attached to the slideway of the machine tool. Fastened to the moving saddle is a much shorter piece of glass having two similar patterns formed on it with a spatial phase separation of one quarter of the pitch of the grid. This slider, as it is called, moves along the fixed scale with an airgap of 0.25 mm. The long grid is fed with a 10 kHz electro-magnetic signal. The slider grids pick up this signal by inductive coupling across the airgap. Phase measurement between the slider elements and the reference oscillator yields direction of movement and position within one gridpitch. It is necessary to determine where the slider is upon the grid and this can be done with a rack driven encoder (which can be of lesser precision), by counting the number of full cycles traversed from a datum position, or by the use of further inductive grids.

This method is basically similar to many other devices – a salient feature is manufactured cyclically along the scale at precisely fixed, constant pitch positions. Position within a cycle is decided by the subdivisional method known as the phase-analogue method, and the number of salient features passed is found by counting, or by reference to a second coarser measuring system. Small magnets, castellations, the thread of the lead screw, inserted slugs – all have been used with magnetic sensing and with the exception of the first, with capacitive sensing also.

Optically-sensed scales – It is also possible to sense marks by optical methods. Opaque lines, small prisms, screw threads and holes have been sensed by the light passing through. The most commonly used method is the first.

In 1950 the British firm of Ferranti Ltd developed a length transducer to facilitate machine-tool control. This employed long diffraction gratings. These were made inexpensively by resin replication from a master unit. The long diffraction grating was used in conjunction with a smaller piece, (called the index grating), to form Moire fringes which were then counted photo-electrically. The ruling pitch was typically 0.001 inch in the imperial measure scale. As time went by the diffraction grating was replaced by an easier to produce and use grid called the bar and space transmission-grating. This simply has nontransparent lines and transparent spaces at the same pitch. The Moire fringes, shown in Figure 6, are formed by placing the index grating at a small angle to the main grating. (These fringes are commonly seen by viewing through the two handrails of a bridge or by looking at a corrugated iron wall through a vertical paling fence). The merit of the Moire fringe is that it is produced as the average of hundreds of individual lines and, therefore, reduces local errors. Furthermore, the fringe pitch is easily arranged to be much wider than the grating pitch enabling large size photo-detectors to be employed.

As with the inductive hairpin grid, two index gratings spaced in phase at 90° give two signals that enable direction to be determined.

The interpolation methods mentioned in the chapter on small displacement sensing were mainly devised to subdivide relatively coarse optical gratings for it is possible to place coarse lines accurately, but impracticable to put more of them closer together. With optical gratings there is no inbuilt time-modulation as with the inductive and capacitive methods. To make use of the phase-analogue method of subdivision, modulation was added by mechanically rotating a reference grating, one method being illustrated in Figure 7.

Later, solid-state methods were evolved. One uses four photocells placed at 90° intervals across the fringe. These are cyclically interrogated and their outputs processed to give phase difference with the cyclic generator. With this technique it is possible to subdivide to one hundredth of a fringe. As subdivision is an absolute measure it is preferable to use a coarse grating and subdivide down to gain the necessary resolution. This reduces errors due to pulse counting loss.

It is also possible to use this method

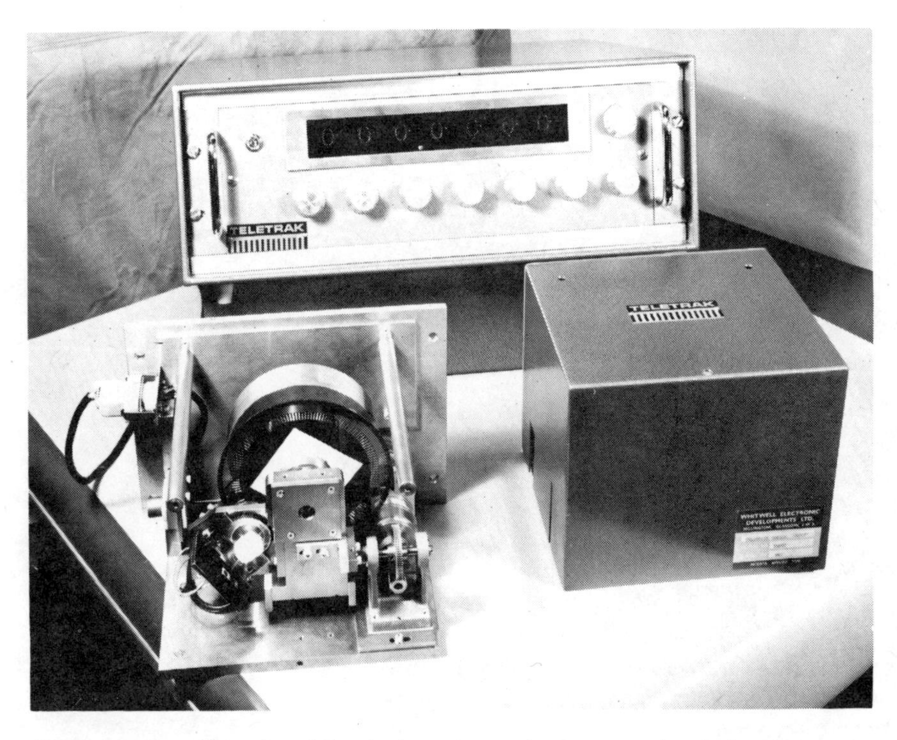

Fig. 8. A commercial version of the absolute system using incremental optical gratings.

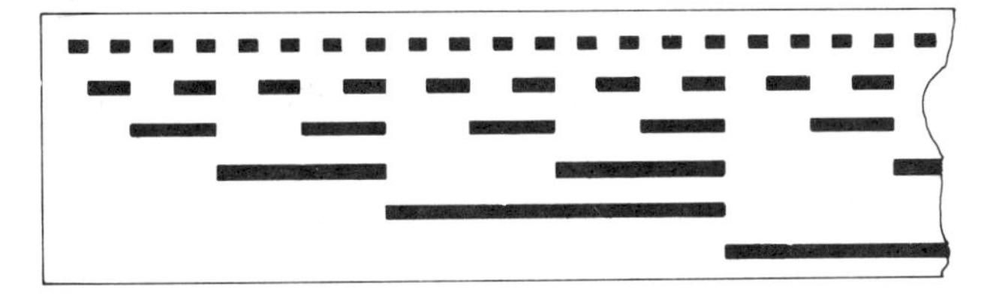

Fig. 9 . Section of a 6 bit absolute digital transmission transducer scale.

Fig. 10. The University of Trieste laser Earth strain meter.

with a reflection grating that has alternate non-reflective, reflective lines etched on a narrow stainless-steel tape. In precision installations the tape is spring tensioned to maintain the same length as when it was etched.

The disadvantage of the simple bar space grating is that it is incremental. An ingenious method has been devised to use them as an absolute system. Consider starting with a coarse grating having a pitch of one millimetre. Phase analogue methods could be used to subdivide the one millimetre cycle into ten absolute parts. If alongside the first grating is one ten times finer, the finer line corresponding to the subdivided position on the coarser grid can be determined by the analogue subdivision. With a third, ten times finer again, grating we can subdivide still further. The position analogue in a millimetre space is found by absolute encoders of low cost. So with incremental gratings it is possible to measure position in an absolute manner. This technique was devised in Britain several years ago and is now marketed by several companies. These successively finer tracks are seen in the angle transducer using this concept which is shown in Figure 8.

This method is tending toward absolute digital optical scales but it needs far less tracks and this eases the manufacturing cost.

Linear digitally coded tracks are available with as many as 20 or more tracks — one with 6 bits is shown in Figure 9 — but linear transducers of this type are difficult to use in practice. However, as we shall see in the next chapter they are used extensively in rotary encoders.

Radiation Scales — As mentioned previously, a laser beam (or any other coherent source of radiation, in fact) can provide a spatial scale if interferometric methods are used. The coherence length of a source decides how far it can radiate and retain a satisfactory wavefront for interference. In the laser this length is such that distances of kilometres are in range. Thus the laser interferometer is especially versatile and is able to measure from millimetres to hundreds of metres with the same apparatus.

In industry, the laser interferometer is too expensive for most routine measurement.

Interferometers are, therefore, usually reserved for vital inspection tasks and for calibrating less accurate scales. They are easily installed — only the laser and a corner cube (or cats-eye) reflector need to be mounted. As the method is by nature, incremental, it is necessary that the reflector is moved in a straight line — within tolerances of a millimetre — to ensure correct operation at all times.

One interesting application of the laser interferometer is its use to control the ruling diamond-carriage position with respect to the moving blank on the Australian C.S.I.R.O. diffraction-grating ruling machine at the Division of Chemical Physics at Clayton in Victoria. Such engines rule, on coated glass, hundreds of thousands of lines side by side and separated by only a micrometre or so. In the C.S.I.R.O. engine, mechanical gears, etc. advance the blank by roughly five fringes. Electro-optic sensing operates a servo that pulls the blank into exactly the fifth wavelength position. A line is then ruled and the process repeated.

Another scientific application of the laser interferometer is for monitoring Earth strains. In these installations the path length over which variations occur, ranges from tens to hundreds of metres. It is the small variations in the length (rarely exceeding a part in a million) that are of interest. The Earth strainmeter built at the University of Trieste is shown in Figure 10. The fixed reference arm and beam splitter are enclosed in the tank. Laser radiation enters through the white tube in the foreground, passes to the far end of the tunnel in the suspended tube, then returns to produce fringes which are monitored by the two photomultipliers mounted on the brick pillar to the left of the picture.

MISCELLANEOUS INDUSTRIAL LENGTH TRANSDUCERS

So far we have been concerned with movements of machines, for this constitutes the majority of industrial range measurements. But there are times when other methods are more appropriate. Let us consider just a few.

Television Gauging — A television camera tube is able to convert an optical image size into an equivalent electrical signal by virtue of a timing process. The vidicon camera tube is the simplest of these tubes and is the most used in gauging applications. The vidicon consists of a target upon which is focused the image of the object of interest. The intensity of the illumination on the target controls the charge distribution on its surface. At the rear of the tube is an electron gun that aims a stream of electrons at the target. This beam is electromagnetically or electrostatically deflected to scan across the target in a systematic manner. The charge on the target decides how much beam current will flow so beam current is a measure of the illumination intensity of the image. Beam current variations (the video signal) and scan position data are combined onto a common signal line and the image is reconstituted in a monitor (if needed).

Fig. 11. This EDM instrument, by Hewlett-Packard, uses a modulated GaAs diode radiation source to measure range. In-built sensors correct for angle and the result is calculated internally.

The tube, therefore, converts image dimensions into time signals, so size of the image, and hence the object of interest, can be ascertained from the video signal. This method is not restricted by object size for the optical system can scale up or down as need be. Applications range from microscope slide examinations, to sizing steel billets in the rolling mill, and tracking of missiles from ground level to that height where radar is effective. Precision is limited to 0.1% of the image size, so it is not in the same accuracy class as machine tool transducers, but the advantages of fast response time and easily adjustable scaling make it attractive in many applications.

Scanned Laser Beams – If a laser beam, which has a divergence of 1 mrad (or less when used with a telescope), is scanned across an object, the time taken for it to reappear after being vignetted by the object is a measure of size. Two ways are employed. In the first the rotation is at a fixed speed so distance is proportional to time of obscurance. Alternatively, the rotation can be locked to an angular resolver in which case scan speed is not critical.

SURVEYING RANGE

There are a number of precise techniques for determining distances greater than a kilometre. These use electromagnetic radiation and are known as electromagnetic distance measuring (EDM) devices.

During World War II both Britain and Germany developed radar to plot positions of friendly and enemy aircraft. The technique was to send a burst of carrier and time its reflected return. Electromagnetic radiation travels at around 300 mm per nanosecond, so the resolution is decided by the ability to detect small time intervals. With early equipments only several metre resolution was possible.

A more accurate way to resolve the time is to use a continuously transmitted carrier and compare the phase of the returned signal with the source. (The phase-analogue method again).

No doubt the development of radar prompted subsequent developments, for in 1949 a device, called a Geodimeter, was announced. In this a modulated light beam is sent out and returned from a distant target reflector. The phase between the transmitted and received signal subdivides the whole cycles and coarse position is found by determining which coarse cycle is being used. As it is not practical to traverse the whole distance and count cycles, a method of frequency changing is used to decide the coarse distance. This light beam method can measure 20 km distances to 10 mm. The latest units, (one is shown in Figure 11), use laser sources to improve the range in daylight.

Then came the Tellurometer – in 1957. This South African development uses microwave radiation. Original units used a 3000 MHz carrier modulated at 10 MHz. Distances from 150 m to 80 km can be measured to the same 10 mm resolution. The latest microwave EDM device operates on a 8 mm wavelength and gives millimetre resolution.

Numerous other devices have followed but with one exception they are modelled after the Geodimeter or Tellurometer. The exception is the Mekometer, a British development from the National Physical Laboratory near London. This uses a short burst of light in which the polarization angle is modulated (rather than the beam angle). By incorporating free air in the modulating cavity, errors due to ambient conditions are reduced. These have been used in land crustal movement research surveys in Iceland and Greenland and to set up the 300 m diameter intersecting storage ring at CERN near Geneva.

In all EDM instruments, the path being measured is in free air, so the pressure, temperature and water vapour content alter the refractive index and hence the travel time of the radiation. These effects limit the accuracy to some two parts per million.

One approach to improving accuracy is to use two systems together each having a different wavelength carrier. Common elements are combined for reasons of economy as shown in the system developed by the United States Environmental Sciences Service Association (see Figure 12). The two measurements can be combined to reduce the path errors by 5 to 10 times.

Until 1968 there was no way to measure short distances (up to a kilometre) with comparative ease. But that year the major surveying instrument manufacturers in Europe each released a ranger which used the solid-state gallium arsenide light emitting diode. This diode is able to provide pulses of intense and highly chromatic light in the red region. In these meters, (one is illustrated in Fig. 13) the transit time technique is used in which the time between sending a pulse and its subsequent return is gated and displayed as distance on a digital readout. Resolution is limited to several millimetres but this is quite suitable for much of surveyors' requirements.

So far we have only considered using phase or gating methods used in conjunction with EM radiation. Acoustic or soundwave radiation travels slower, by a factor of at least one million, so better resolution is possible for a given technological limit on gating a pulse.

Sonar and Asdic are devices for

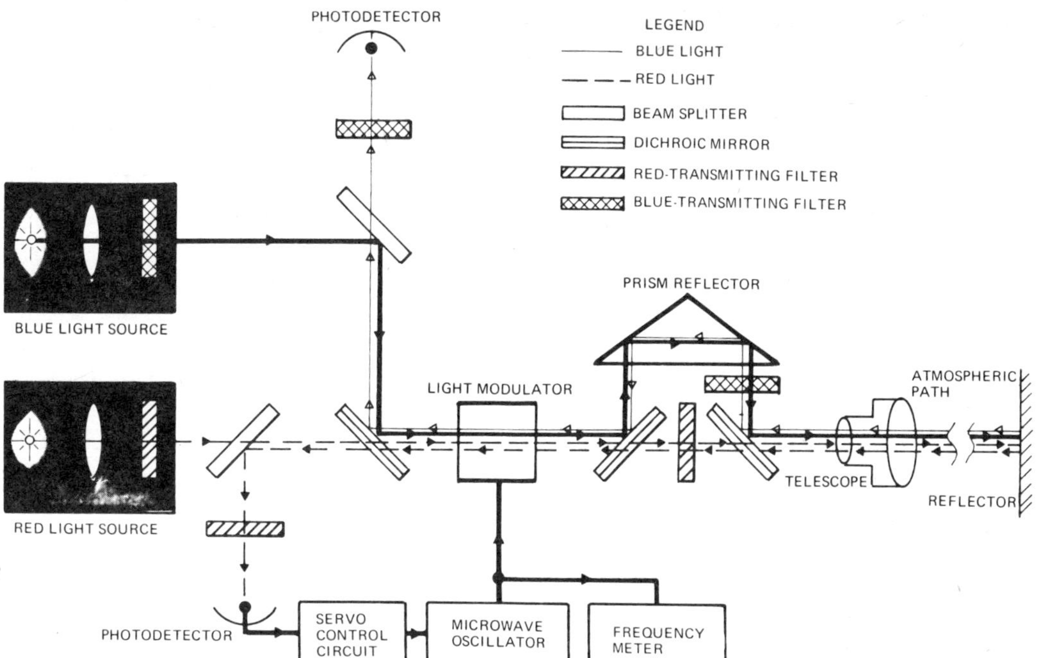

Fig. 12. A schematic of a two-colour distance measuring technique used by ESSA in the United States of America.

Fig. 13. Gallium-arsenide diode powered transit time ranger.

ranging under water. Ultrasonic methods work well in a dense field or a solid as the transducers can be efficiently coupled to the medium transferring the wave. Ultrasonic methods are less accurate in air as the coupling is poor.

Ultrasonic distance gauging is applicable for both long and short ranges (as mentioned in the last chapter). It has been used in submarine tracking, surgical probe guidance, water level sensing, determination of wheat level in silos and thickness measuring in industry. An interesting application is its use to guide a deep sea drill bit into an entry cone fastened on the sea floor many thousands of metres below a drilling platform.

CELESTIAL RANGE

As man's desire to explore space is realised, it is increasingly necessary to know where an object is with respect to another planet — for instance the Earth. Better measurements of distance are possible in space due to the absence of an atmosphere, and also because the distances are large enough to enable transit-time methods to be effectively used, for gating error is insignificant.

During the recent Moon visits, retro-reflectors were placed on the surface to return powerful pulsed laser beams back to Earth. From this method it has been possible to determine the distance to the Moon to an incredible accuracy, and we are now able to observe such effects as the Chandler wobble of the Earth by monitoring the Earth/Moon distance variations.

A space vessel is guided to another planet by servo-systems locking onto the planets image, or by relation to certain chosen stars in the star field. Once near to the planet the vessel locks on to the horizon of the planet until its radar is within range. For the last few hundred metres of descent it might use ultrasonics, television or even a weighted tensioned cord to control the final approach velocity. These have all been used at some time or another.

So far we have described how we can transduce lengths from 10^{-14}m to 10^{20}m or more into electrical signals.

In the following chapter, transducers for converting angles into electrical signals are outlined — these are very much tied to the measurement of dimension and position.

FURTHER READING

"Linear and Angular Transducers for Positional Control in the Decametre Range". P. H. Sydenham, Proc. IEE., 1968, 115, 7, 1057-1066.

"Transducers for Positional Measuring systems". R. C. Brewer, Proc. IEE., 1963, 110, 10, 1818-1828.

"Plane and Geodetic Surveying — Vol II. D. Clark, Constable, 1963.

"Electromagnetic Distance Measuring — EDM". Hilger and Watts, 1967.

"Ultrasonic Engineering". J. R. Frederick, Wiley, 1965.

"Instrument Transducers", HKP Neubert, Pergamon, 1975.

"Interferometry as a Measuring Tool", J. Dyson, Machinery Pub. Co, 1970.

"Handbook of Applied Instrumentation", D.M. Considine and S.D. Ross, McGraw-Hill, 1964.

"Microdisplacement Transducers", P.H. Sydenham, J. Phys. E.: Sci. Instrum, 1972.

"Position-sensitive Photocells and their application to static and dynamic Dimensional Metrology", Optica Acta, 1969, 16, 3, 377-389.

CHAPTER 3
AUTOMATIC ANGLE MEASUREMENTS

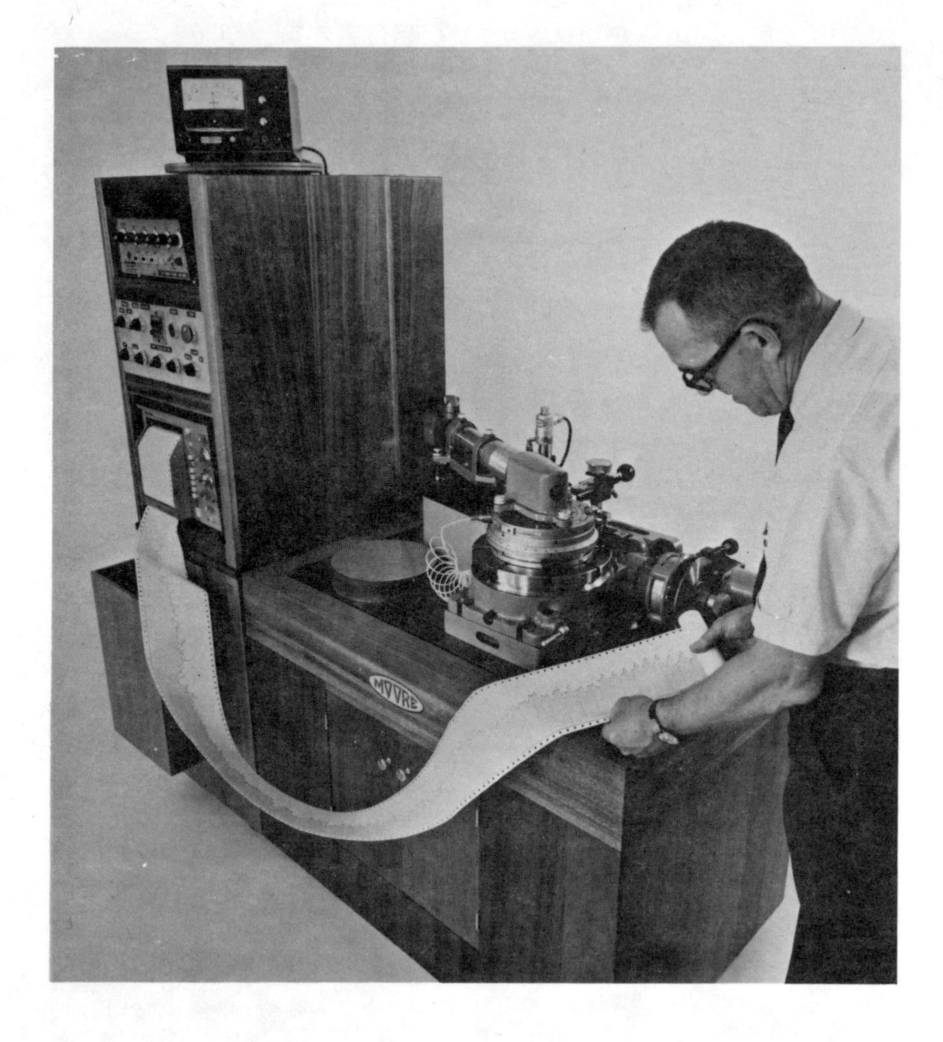

Calibration of precision turntable at Moore Special Tool Co, USA.

IN the previous chapters we have covered how lengths, ranging from microscopic to macroscopic, can be transduced into electronic signals using electrical, mechanical and optical techniques. We have discussed this parameter thoroughly because length is a fundamental parameter often leading indirectly to the measurement of other secondary variables. In fact it has been said that some 85% of all measurements involve length.

Another very common transducer group includes devices that convert rotation into the common electronic language with which technical systems are interconnected. Angle transducers go under a variety of names – the terms, resolvers, encoders, angle transducers, rotary transducers, digitisers, shaft position indicators and synchros, all are used depending upon the principle employed. Often the names are applied synonymously – and incorrectly.

Angle is dimensionless. It is defined as subdivision of a circle, in a number of ways. Commonly used units are the degree and the radian. There are 360 degrees in a circle (and 2π radians). The latter comes about by defining radian measure as the ratio of the arc length to radius for a segment of a circle. Radians are easily used when small angles are involved for they approximately express the deviation per unit distance, for example, a milliradian is a divergence of 1 part in 1000. There is another system, used extensively in Europe, which has 400 divisions, called grades, in a full circle. This gives a convenient 100 grades per quadrant. Grades are subdivided into centigrades and centi-centi-grades, i.e., subdivided in hundreds compared with sixties in the degree-minute-second system.

As angle is dimensionless, it is defined completely by a mathematical expression. No physical standards are needed from a theoretical viewpoint. (Compare this with length for instance, which uses quite arbitrary units, chosen by man to suit his convenience). However, it is more practicable to maintain angular standards in the form of divided circles, angle gauges and optically-worked multi-sided polygons made of metal or glass. A commercial

Fig. 1. Test set up used by Data Technology. The resolver is mounted underneath the plate holding the micrometer. On top is a standard angle polygon which enables the resolver to be rotated in exactly known angular steps. The autocollimator on the right is used to set the rotation by an optical lever action on the polished flats of the polygon.

test set is shown in Figure 1 where a polygon is being used to check a resolver system.

As angle is defined in terms of two lengths, it is clear that angle transducers generally use devices that measure displacements by fixing one length and measuring the other as it varies. There is, however, one important difference, for angular transducers can use mechanical components of rotation that can be manufactured more precisely than linear components for equal costs. This enables closer tolerances to be maintained between two measuring surfaces. Also it enables the simple incorporation of both spatial and time averaging to improve the precision of the device. These features will become clearer later.

As with length measurement, there is a distinct difference between the technique that may be used to measure small angles of less than say a few degrees and those required to cover greater angular excursions. A broad assumption is that the smaller the angle to be measured the higher the precision that can be obtained. At one end of the scale are devices capable of measuring 10^{-10} radians over a range of 10^{-6} radians (about an arc second). At the other are devices capable of arc second resolution with a continuous full circle range.

So much for a general background to angle measurements. Let us now consider the techniques that have proven practicable.

SMALL RANGE ANGLE TRANSDUCERS

The most obvious means by which a small angle can be monitored is to measure the linear displacement of the free end of a hinged arm of fixed length using one of the length transducers described earlier. This concept has been used in the force-balance principle, see Figure 2, used mainly in process control and weighing balances. In this a beam is hinged in the centre. The force to be measured is applied at one end of the see-saw beam. At the other end is a magnetic solenoid to which variable current is applied in order to balance the unknown force. The current in the coil at balance is then a measure of the applied force.

The beam is considered balanced when it is at reference position. To establish when balance is achieved, the small rotations of the beam must be transduced. Virtually all fine displacement techniques have been used to monitor the displacement of the beam, ranging from simple on-off contacts to devices providing a proportional output.

In some applications, such as torsion balance and galvanometer readouts, it is not permissible to load the driving force with a relatively massive lever that would be needed to operate a transducer. In these cases the optical-lever comes into its own.

The optical-lever is generally attributed to Poggendorf who described it in the literature in 1826. No doubt it was used even before that. In the days before electronics, that is before this century, the optical-lever was the only sensitive way to sense small angular movements and displacements. The principle is simple. A source of light, collimated to a near-parallel beam, is reflected from a mirror surface mounted longitudinally on the axis of rotation to be monitored. In the purely mechanical use of the lever the reflected beam impinges on the rear of a transparent graduated scale. The sensitivity of the method derives from the optical phenomena of angle-doubling at the mirror and from the ability to place the limited-resolution graduated scale at a considerable distance from the mirror.

In most modern applications, however, the aim is to provide an

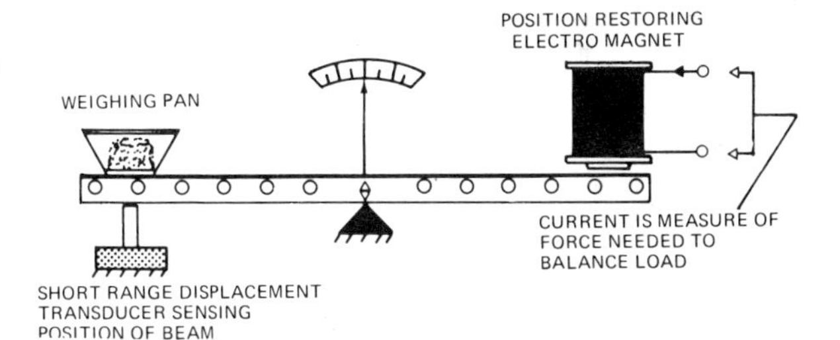

Fig. 2. The principle of the force balance.

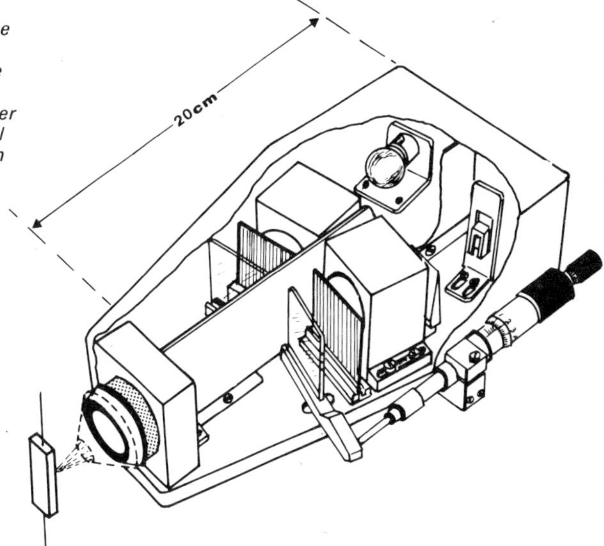

Fig. 3. Construction of the 'Jones' optical lever. Two grids are used as bar-space gratings to enhance the sensitivity. The micrometer rotates the optical parallel plate to displace the beam for fine adjustment.

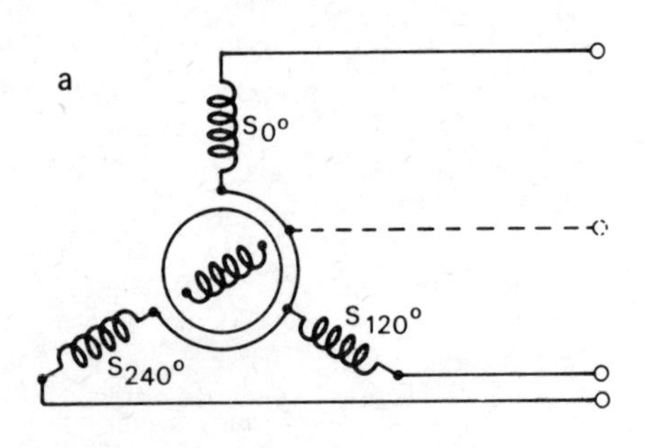

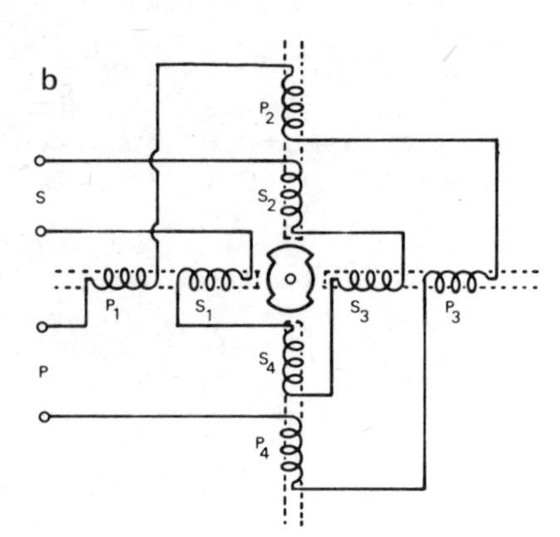

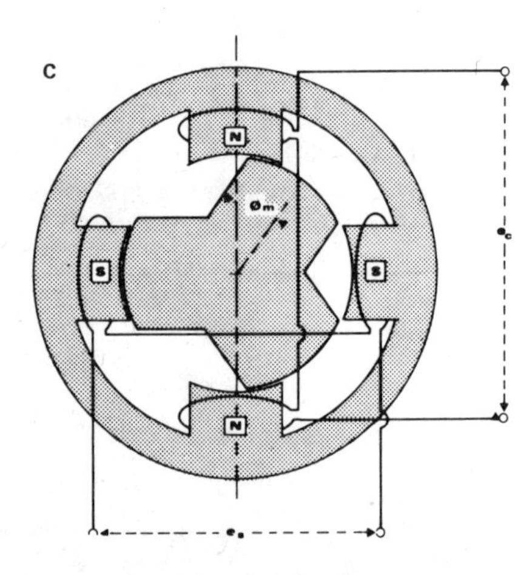

Fig. 4. Inductive rotary transmitters. a. The Synchro b. Reluctance resolver c. Vernier resolver (excitation windings are also on each pole).

electrical recording readout and keep the size of the angle transducer down to small dimensions. In 1920, Wilson and Epps proposed that two thermopiles be used to detect the position of the light spot instead of relying on the relatively poor resolution of the eye. (Thermopiles are a bank of thermoelectric elements cascaded in series to form a sensitive heat-sensing device). If the two thermopiles are connected in opposition, the resultant current flow would be a measure of the spot position upon them. Such a system was soon constructed and the first

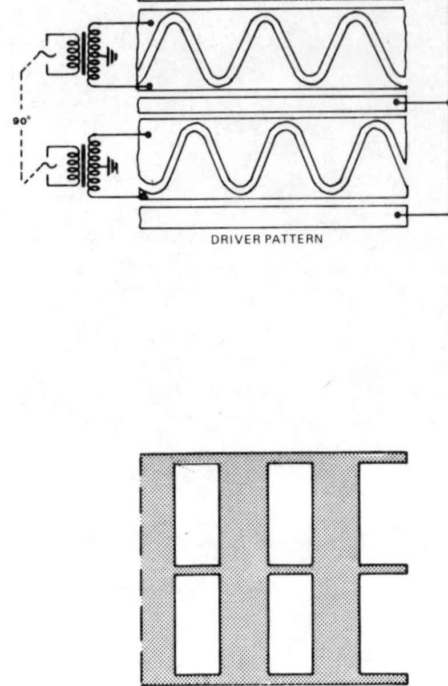

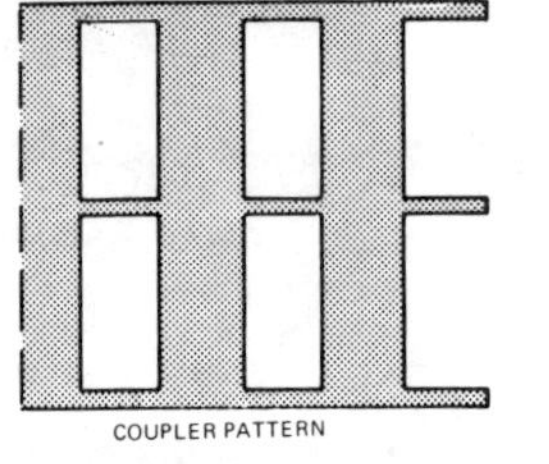

Fig. 5. Section of metallic pattern photo etched on glass supporting disk. The coupler pattern runs close to, but not touching the driver pattern, forming a capacitive bridge network.

ultra-sensitive angle sensor was created. It was so sensitive, in fact, that it could detect the Brownian movement of the mirror suspension. Then came photo-electric cells which were able to detect much smaller quantities of light energy. A great exponent and developer of the optical lever is Professor Jones and his colleagues of Scotland's Aberdeen University. They have built extremely elegant lever systems that can detect angular excursions of around 10^{-10} radians. One of these is shown in Figure 3. With these, Professor Jones has studied many interesting physical phenomena, including measurement of the radiation pressure of a liquid at atmospheric pressure (this is much harder than measuring it in a vacuum), determination of the length changes of an X-ray irradiated lithium-fluoride crystal, observation of an I-R energy level of 10^{-11}W, and verification that a light beam is not slowed by a magnetic field. The latter has been done to an extraordinary degree of precision. More recently Professor Jones has used the optical lever to verify the Fresnel drag effect, (predicted by Fresnel many years ago).

Not all applications need such exacting stability and precision, so most optical levers are no more than a lamp, a collimating lens with spot aperture built in, a mirror, and a dual photo-cell position-sensitive detector. This simple arrangement has been used to provide electrical readout from a bourdon-tube microbarograph, from sensitive galvanometers, recording autocollimators (used in industry for alignment – to be described in our next chapter) and in sensitive microbalances, in other words, in applications where the angular excursion is small and the driving force cannot tolerate any load being imposed upon it by the transducer. The main disadvantage of the optical lever is that the light-source generates a considerable amount of heat and this may cause error due to thermal expansion effects. The ever-reducing price of continuous wave laser sources makes them attractive, for the beam is as well collimated as would normally be required. A divergence of 1 mrad is typical.

Alignment devices for measuring deviation from a straight line involve the measurement of small angles. This class of transducers, however, deserve a chapter to themselves as they are extensively developed. The reader is, therefore, referred to the next chapter for many details of small angle measurement over long distances.

LARGE RANGES

It can be seen that the use of optical lever (or microdisplacement transducer) angle measurements cannot cover more than a few degrees of rotation unless mechanical angle amplifiers (gears, belts, friction wheels) are used. Other methods, therefore, have been devised to cope with larger excursions.

Early this century, divided circular scales could be made to a precision of better than one arc second. Geodetic-grade theodolites have glass scales of 100 mm diameter. These can be read to around 0.1 arc second (that is, the circle is divisible into 12 960 000 parts!) using a reasonably simple optical viewing system.

The majority of wide-range angle transducers can be grouped into those having mechanical movement no faster than their input slewing rate, and those in which continuous rotation at a speed higher than the slewing rate

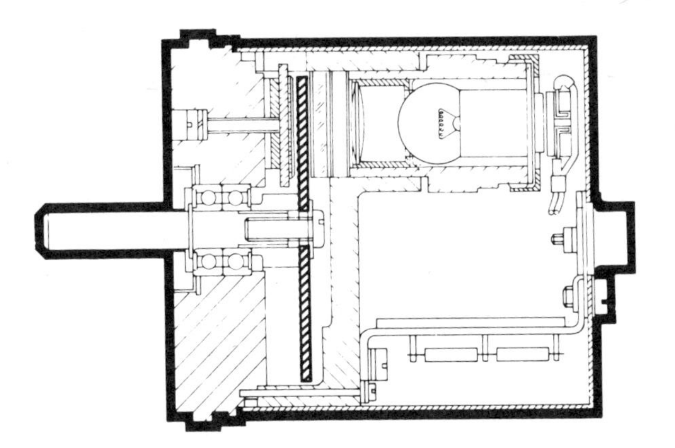

Fig. 6. Cross sectional view of optical shaft encoder.

has been deliberately added. The reason for the latter will become clearer as we proceed. For simplicity these will be referred to as static and active systems.

RESISTANCE POTENTIOMETERS

Large-range angle is very simply measured with rotary potentiometers provided the contact friction and wear disadvantages can be tolerated. But it is difficult to produce a full 360°, continuously rotatable unit, as the wiper shorts the start and finish of the resistance track unless a dead zone is used. On top of this problem is the very rapid change of signal level at the end. Multiturn units are used if the order of a few complete rotations is needed but these cannot cater for infinite rotation.

The problem can be overcome by using a complete circle resistance track in which the end is joined to the start. If a´ supply is connected at diametrically opposite points, the track forms a bridge network. A single wiper will produce a signal varying from positive to negative without any abrupt signal level change. It is necessary, however, to add a system that can recognize which half the slider is on, for there are two positions giving the same output.

INDUCTIVE AND CAPACITIVE STATIC METHODS

A similar technique can be used with a toroidal transformer but again brushes are needed. Avoiding the need for sliding contacts leads logically to the inductive synchro (also known as a magslip transmitter). Inductive resolvers are in this class being only slightly different in construction. The synchro consists of an armature made to be a simple rotating bar magnet which is fed with ac alternating current – via slip rings – see Figure 4a. The stator houses three, equally phase-shifted identical windings, similar to those found in a three phase induction motor. The ac excitation in the armature induces voltages in each of the rotor windings which are phase shifted with respect to each other. At any given position the amplitudes of the outputs from the stator are unique. Position is thus defined in an absolute sense. By feeding these signals to a similar synchro receiver the position can be reproduced remotely (hence the name transmitter). This method still uses brushes but these are not in the measurement circuit. They are there to provide high-level ac energy to the system. Better quality synchros can resolve around 1 minute of arc – this is adequate for many tasks. To obtain greater precision, accurate gear trains may be used, and the designer has to decide whether a synchro combined with expensive gears is a better economic choice than the use of more sensitive direct methods. If the object whose rotation is to be measured is large, for instance a steerable radio telescope, the physical size of the transducer is of no importance and gears or a large diameter resolver can be employed. If a small size is essential then tea-cup size devices are needed. The need for brushes is avoided if excitation is provided by windings placed on the stator pole-pieces using the rotor to vary the reluctance and thus vary induced voltages, (as shown in Figure 4b).

Before leaving the synchro it is worth mentioning a development in 1956 which uses the variable reluctance transmitter combined with the Vernier principle. The latter effect (named after Pierre Vernier who lived at the turn of the 17th century) is a way to subdivide the minor divisions of a scale using a similar sized scale but having, say, 10 divisions when the main scale has 9. Whatever the position of the Vernier scale there will be one line upon it that is aligned with the main scale division. The Vernier scale line in coincidence gives the subdivision of the scale interval. In the Vernier resolver, shown diagrammatically in Figure 4c, the ac-fed armature has one less pole than there are stator poles. Processing of stator outputs enables the position within each electrical cycle to be resolved more finely than with a standard synchro. In units built with 33 rotor teeth a repeatibility of ±3 arc seconds was obtained.

One feature that gives the Vernier resolver high precision is that the output signals are formed as the average of inductive coupling between many iron circuits around the multipole core. This is a powerful measurement principle and is termed spatial averaging. It also relaxes the centring tolerances needed for a given angular accuracy. Being an ac method, a measurement is available with each cycle of the driving frequency. In this case the system uses 400 Hz. If the time response can be lengthened, the output can be time-averaged to gain further improvement again. A general law of errors states that such averaging processes (if they involve random errors) improve the precision as the square root of number of measurements involved, so in this case a hundred-fold improvement is theoretically possible for a 1 Hz bandwidth and about 30 poles.

Whereas time-averaging can be incorporated into linear transducer systems, complete spatial averaging over the whole measurement device usually cannot. For this reason angle-transducers can be quite small for a given performance. But more of this when active methods are discussed.

It can be readily seen that the more poles there are, the finer the resolution, so in cases where space permits, a circular form of the linear inductive, or capacitive, grating can be used in which radial, rather than linear, fixed plated windings are inductively coupled with a rotating

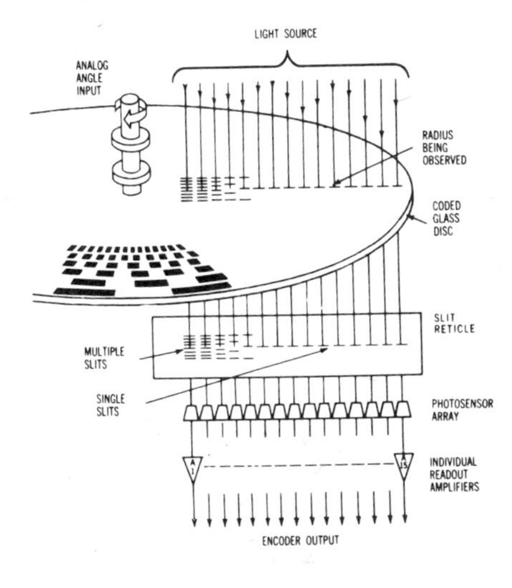

Fig. 7. The multiple slit optical encoder system used by Computer Control Company.

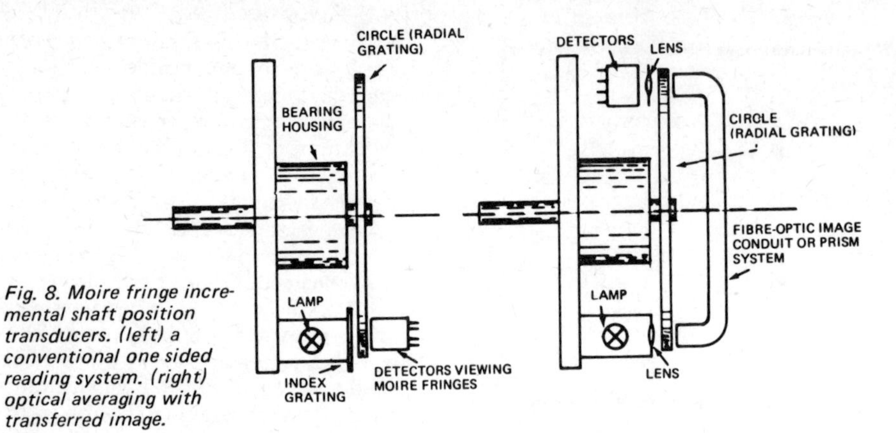

Fig. 8. Moire fringe incremental shaft position transducers. (left) a conventional one sided reading system. (right) optical averaging with transferred image.

pickup plate. Radial systems have the distinctive advantage that they can be manufactured cheaply by contact photo-mechanical methods whereas cylindrical devices must be individually machined when high resolution is needed.

Virtually all that has been said for inductive systems can be applied to capacitive coupling but, in the main, only the pancake radial method has been adopted for continuous rotation measurement. Figure 5 shows the construction of this.

STATIC METHOD – OPTICAL

Although the earliest angle measuring machine was the divided scale, it was not automated until the late 1950s. As inductive synchros could not provide arc second resolution at that time, attempts were made to read ruled scales automatically.

Mechanical contacts were far too large so it was logical to use optical sensing as this was the method by which the then available scales were read. At that time there was also a growing need in digital computers. One difficulty encountered with the design of fine resolution optical absolute-encoded binary disk systems (similar to the absolute digital length scale mentioned previously) is that the apertures of the lower significant-digit tracks are extremely fine, thus restricting the amount of light transmitted. A cross sectional view of a low cost encoder is given in Figure 6. Although expensive, absolute encoders are now available with around one arc second resolution. In these, discharge flash lamps are used to illuminate photo-diodes positioned on the opposite side in order to read the code. One unit, shown in Figure 7, uses multiple slits to ease the illumination difficulty. It is, however, far easier to make a 21 bit encoder in rotary form than it is to produce satisfactorily in linear form as the disk can be made to rotate with adequately constant clearance more simply than in the latter case.

It is also quite practicable to use the linear Moire-fringe principle, originally devised as a length measuring method, in a rotary form. Numerous Moire-fringe shaft digitizers are available commercially. The optical gratings have radial lines produced around a disk by photographic or photomechanical methods. Full circle radial gratings having over 30,000 lines have been made, the main institutions responsible for their development being the National Physical Laboratory (NPL) and the National Engineering Laboratory (NEL) in Britain. The absolute grating method, using incremental gratings, is also available in rotary form (see previous chapter for illustration).

In the Moire fringe method, (Figure 8), counting is necessary to determine angular movement, and hybrid systems have been marketed using phase-analogue subdivision of relatively coarse gratings. The National Electrical Manufacturers Association of America (NEMA) has standardised the name of these and other incremented methods, as rotary incremental digital position transducers (RIDPT for short).

Theodolite scales are read to the highest precision by viewing diametrically opposite values simultaneously. This averages the bearing centring error. The idea has been used in some Moire-fringe resolvers by forming fringes at one radial position of the disk using the other side of the disk as an index grating. A coherent fibre-optic bundle, Figure 8, or a prism system, transfers the image. Still further improvement is possible if the signals from four reading heads placed at 90° around the grating are averaged and this has been used at the NEL and other places.

MECHANICALLY ACTIVE SYSTEMS

Taking it to the limit, the ideal therefore, is to have an infinite number of reading heads placed around the grating to produce complete spatial averaging. Practical limits are set on this idea by the accuracy of the grating lines, for spatial grating errors must lie within certain limits if the signals are to be useful in practice. Bearing eccentricity also enters the picture, for slop or out of roundness of the bearing support system will allow the main grating to move, radically changing the shape and pitch of Moire fringes quite considerably. This can be overcome by adding a continuous rotation to the system, a method that evolved first as a capacitive device.

In 1957 Richard Webb of the U.S.A., filed a patent application for a then quite unusual angle encoder. The device is shown diagrammatically in Fig. 9. A reasonably constant speed

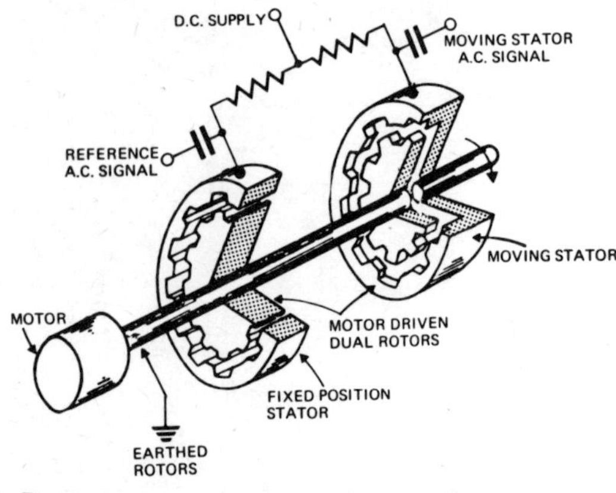

Fig. 9. The capacitive dual generator angle transducer.

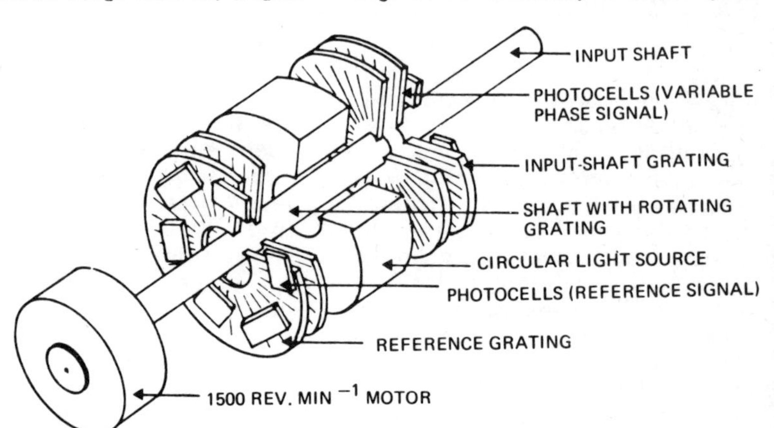

Fig. 10. Optical equivalent of capacitive system in Figure 9.

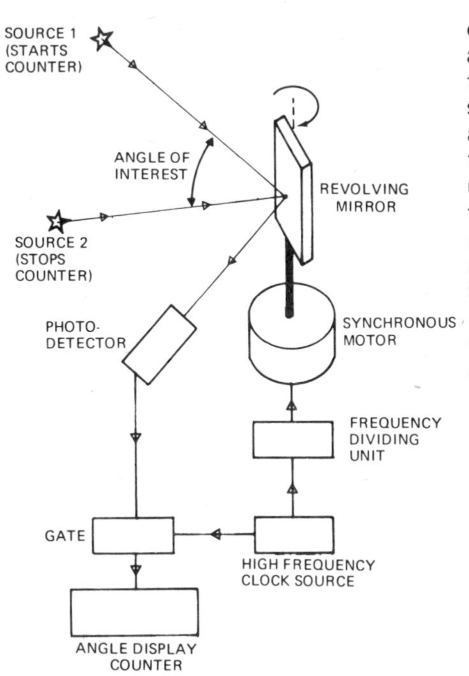

Fig. 11. Chronometric method of measuring angle.

motor drives two toothed rotors each spinning inside toothed stators. One stator is fixed, the other rotates as it is connected to the input shaft whose rotation is to be measured. As the stators turn, the capacitances between the electrically insulated stators and the grounded rotors vary. This is used to produce two ac output signals. If the input-shaft rotor is rotated with respect to the fixed one, the phase of its generated signal varies with respect to the other. Hence the phase difference between the two signals is proportional to the extent of rotation within one pitch of the stator teeth. This is in reality the phase-analogue technique and to complete the system, a coarse angle resolver is needed to form a complete hybrid arrangement.

The original device containing both fine and coarse resolvers was only 50 mm in diameter — but a rack of electronics was needed to interpret the device. To obtain direct degree readout, 360 teeth were cut. As each cycle could be subdivided into 1000 parts by electrical phase methods the transducer could resolve 3.6 arc seconds. The rotors were spun at 900 rpm. This method, therefore, incorporates both spatial and time averaging to advantage. The principle was subsequently used with inductive coupling and the makers of one device claimed 0.1 arc second resolution from a unit 125 mm in diameter.

It was obvious that an optical equivalent existed to these, and in 1967 an experimental unit using rotating radial gratings was made (Fig. 10). This crude unit used low quality 360 line gratings made on plastic sheet. The accuracy of subdivision over one degree of rotation was tested and found to be ten times better than the known errors of the grating, thus showing the power of incorporating averaging. The advantage of the optical form is that optical gratings can be used with a far higher density of lines than magnetic or capacitive as the light rays are not as subject to fringing errors when coupling the two grids. Secondly, optical gratings have been extensively developed and are available with 1 arc second accuracy in 2 inch disks. Finally, but by no means the least factor, radial gratings can be copied inexpensively (as has been mentioned above). It seems that a well made optical dual-modulator angle transducer might realise 0.01 arc second accuracy using normal precision ball bearings instead of needing the hydrostatic systems that have been used in extreme precision experiments.

CHRONOMETRIC METHODS

Another interesting method for converting angular rotation into electrical signals is known as the chronometric method by which angle is transformed into time.

Consider a shaft being driven by a synchronous motor which is energized by a divided-down higher clock frequency (as shown in Fig. 11). If on the shaft is a trigger mechanism that responds to stationary objects (these could be two distant signal lights placed at different locations and for which the subtended angle is needed), then this trigger can be used to gate the clock source into a counter. As the speed is synchronously related to time, the angle can be determined in terms of time, to high accuracy. Time averaging could be added if the system were driven fast enough.

This chronometric system is akin to another angle dividing and measuring system. A magnetic wheel is driven with a synchronous motor and the periphery of the wheel is magnetized from the same ac signal forming magnetic zones around the wheel which are exact integers of the circumference. It is then used to measure position using sensing heads.

The full potential of chronometric methods has not been realised. Nano-second rise-time signals are now commonplace and a shaft driven at 600 rpm could probably be resolved into 10^8 parts — but that remains to be seen.

INTERFEROMETRIC METHODS

Interferometers can resolve extremely small distances and yet have an enormous range. Attempts have been made to measure angles with interferometers and one device capable of 0.01 arc second resolution over a 30° arc has been described. A schematic of this method is given in Fig. 12. A light source, such as a spectral lamp or laser, is used to illuminate the interferometer. As the reflecting corner cubes rotate together, the relative lengths of the two arms change and the fringes move.

Another recently devised laser method that shows promise is the ring laser. Instead of the laser having the usual linear cavity made between two end mirrors, it has a 'ring' cavity. The simplest approximation to the ideal ring is a triangular system. One feature of the cavity is that it can support independent oscillations in both the clockwise and anticlockwise directions. If the ring is rotated about an axis perpendicular to its plane, the two oscillations vary in frequency and upon square-law device-mixing a beat-frequency is produced that is dependent upon the velocity of rotation. The ring laser has been developed mainly for gyroscope applications but an angle transducer version is possible if use is made of this velocity signal output. One advantage of the method is that the centre of rotation is left clear and the ring can be built around a central object.

In the next chapter we will consider ways to transduce a level line of sight or a vertical plumb-line and how to determine alignment along a line. It will then be possible to discuss how we combine these with angle and length transducers in order to determine position on a plane or in a three dimensional space.

FURTHER READING

''Some developments and applications of the optical lever'', R. V. Jones, J. Sci. Instrum., 1961, 38, 37-45.

"Linear and angular transducers for positional control in the decametre range", P. H. Sydenham, Proc. IEE, 1968, 115, 7, 1057-1066.

"The design of optical digital instruments", I. R. Young, Electronic Engineering, 1960, 132, 388, 359-365.

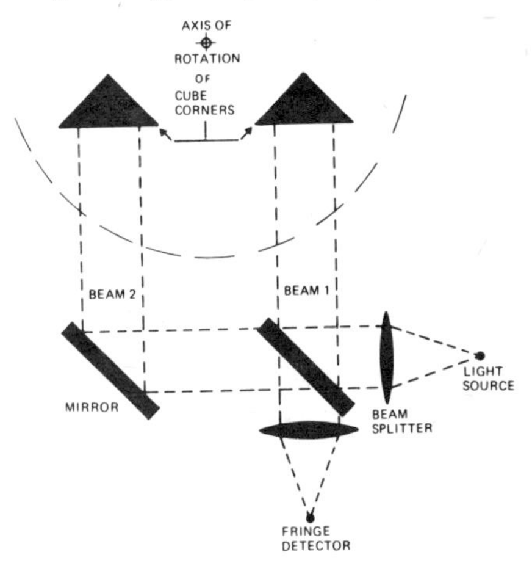

Fig. 12. Differential-length angle measuring interferometer.

CHAPTER 4
TILT AND ALIGNMENT

Laser alignment system for construction work. (Laser Electronics)

The force of gravity causes liquids to settle with a horizontal surface – and suspended objects to hang in a vertical direction. These two natural reference directions are used extensively in engineering construction, e.g., in the erection of buildings and bridges; in agriculture where drainage is vital; in road and railroad building to obtain smooth curves, and in the workshop when flatness or straightness is needed. The plumb line provides a perpendicular to the horizontal plane so each may be derived from the other.

The Earth, being roughly spherical, has a curved level surface with the verticals being at different angles to each other at different locations. For most engineering structural requirements, however, it is adequate to regard the area of surface involved as flat. This curvature is roughly one part in 300 000 (0.1 mm in a 30 m distance) and this is only relevant in the construction of the most precise engineering structures, such as large nuclear accelerators.

If these phenomena are studied more closely it will be found that the liquid surface does not smoothly vary around the Earth in a spherical shape but takes up an undulating surface. This is the result of the varying gravitational forces brought about by the different distribution of mass in the Earth. The surface varies periodically in direction by a small amount, this being the result of the influence of the Sun and Moon which cause shape changes in the Earth. In geophysics these changes in the level surface or the vertical are monitored with great sensitivity in order to study the behaviour and composition of our globe. There are, therefore, many disciplines needing devices that can produce an electrical signal when deviations from the horizontal or vertical occur. In general, engineering inclinations need only be resolved to around an arc second at the best (but with a dynamic range of degrees), whereas in geophysical measurements the need can be for the utmost in resolution with a range rarely exceeding arc seconds.

Another group of closely allied devices are those for measuring alignment. As many of these have levelling devices inbuilt, it is appropriate to discuss them together. An alignment device is capable of yielding measurement information about the degree of displacement of a point from a chosen line or plane surface, but usually there is no provision for deciding where that point is along the line.

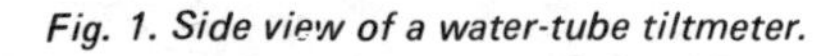

Fig. 1. Side view of a water-tube tiltmeter.

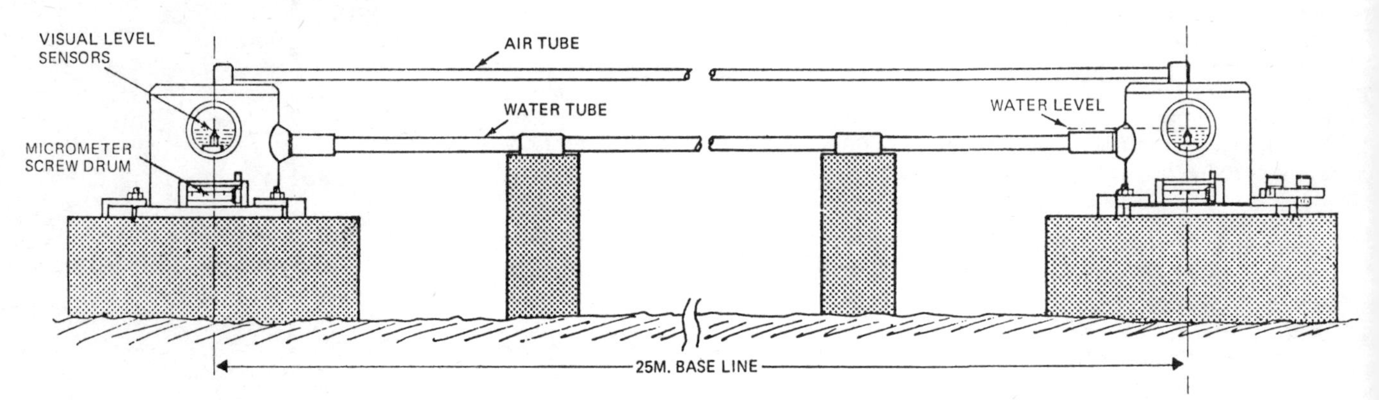

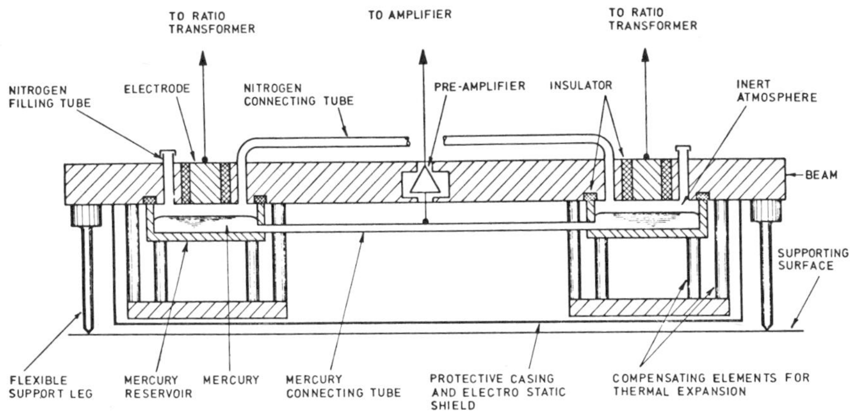

Fig. 2. Cross-section of a mercury cistern tiltmeter.

TILT TRANSDUCERS

Liquid Level References

A large percentage of tilt transducers make use of water or mercury pools which are interconnected with a communicating tube. As tilt occurs the pool heights at each end vary relative to their container. The changes in height are monitored with microdisplacement devices such as are described in Chapter 1.

Tilt is basically an angular measurement, so it is apparent that the further the containers are apart, the greater the displacements resulting. A schematic view of one of a few water-tube tiltmeters installed in New Guinea for crustal movement research is shown in Figure 1. This unit does not have automatic recording but relies upon visual observation of a needle that is manually driven upward with a calibrated micrometer-screw — to a position where its point just breaks the surface. Using the microscope viewing units provided to see when the surface is broken, it is possible to resolve a 3 μm difference in height in the 25 m base line used, giving it angular discrimination of 10^{-7} radians.

Another application of the water-tube method is for monitoring the settlement of the structures of generating plants in power stations. It is not convenient to climb over the plant in order to read the individual levels. To avoid this, a system is used that enables the operator at a central point to pump up the level at each remote container in turn, until an electrical circuit is made by the liquid touching an electrode. He then reads off the level at that point from his console.

The leaning tower of Pisa is instrumented with a liquid circuit around its base. Transducers operated by floats give the tilt of the tower relative to the horizontal datum provided by the liquid. Diametrically opposite gauges provide differential signals that reduce errors due to level changes as the liquid heats and cools or evaporates.

The largest type of liquid level measurement must be the sea-tide gauge. A common method uses a float driving a rotary transducer via a chain or wire. The units act with less than unity gain, for the amplitude of the movement is large. To obtain a well-conditioned response from the float, it has been found necessary to use a hydraulic filter consisting of a vertical tube containing the float with small entrance and exit holes that damp the rate at which the water can enter or leave the tube. This acts as a low pass filter removing the high frequency components.

The main difficulty with an extreme sensitivity water system is how to sense the surface position. If greater resolution were available in the surface detection, the base-line could be shortened reducing the size of the equipment. By using a conducting liquid, the liquid itself can act as a common electrode in a differential capacitance sensing arrangement (see Chapter 1). For this reason several mercury-cistern tiltmeters have been developed that have extreme sensitivity with only centimetre baselines. A cross-section of the ANAC instrument originally developed at the University of Queensland, is shown in Figure 2. Above each mercury pool surface is an insulated electrode; these and the mercury form part of a bridge circuit which is completed by an electronic unit using ratio transformers. The unit, illustrated in Figure 3, can measure angular changes as small as 10^{-9} radians. This is two orders of magnitude better than the much longer water tube tiltmeter described above.

Not all tilt has to be measured with such exactitude. In building construction, for instance, the requirement is for only millimetre definition in metre distances. The familiar spirit level is the oldest form of liquid level in general use. In this a gently upturned curved vial contains a liquid in which a small air bubble is trapped; the bubble attempts to remain on the top of the curve. Sensitivity increases as the curvature flattens, and a good quality engineer's level can discriminate tilts from the horizontal of micrometres in a metre.

The bubble level has been automated by British Aerospace. This unit has platinum electrodes set into the glass and the unit is filled with a conducting alcohol solution. As the bubble moves in the vial, the electrical resistance between the central and outer electrodes varies and the movement can be sensed using an ac bridge as shown in Figure 4. The most sensitive version can sense fractional seconds of arc with a settling time of less than a second.

Occasionally there is need for two-axis level readout and for this the bubble method has been employed by NASA personnel in a different form. A circular bubble forms a lens that modifies the distribution of light passing through it. A reflecting mirror is placed above the bubble. Four photocells, used as position-sensitive

Fig. 3. Mercury-cistern tiltmeter with electronic unit.

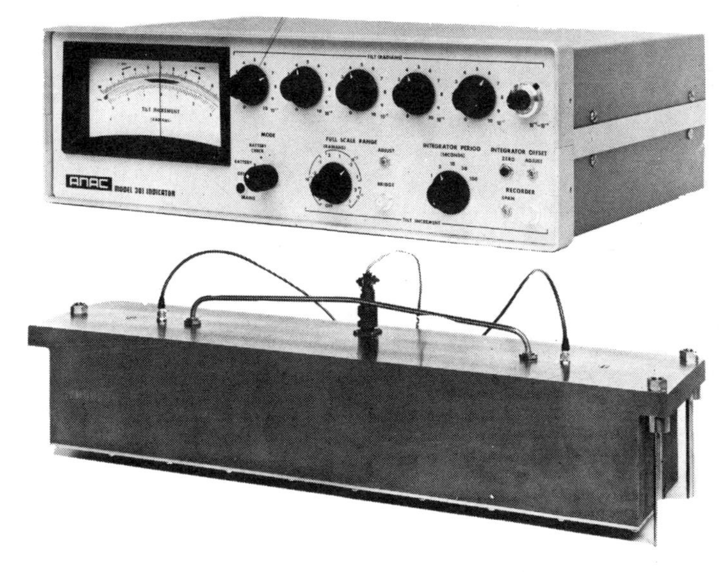

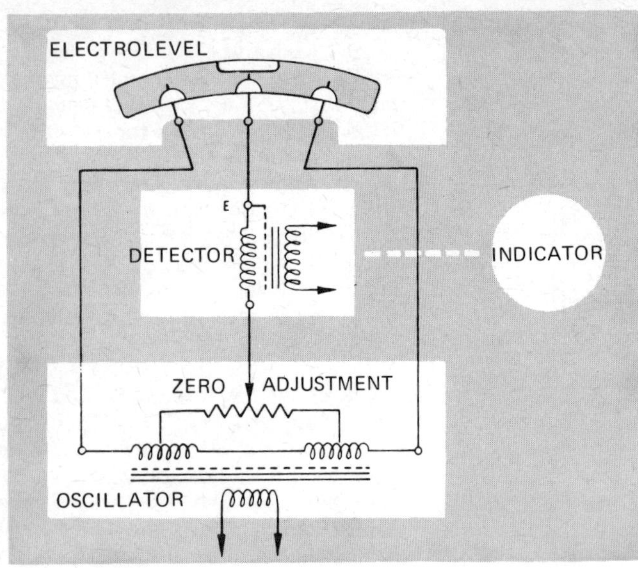

Fig. 4. Electrolevel tilt transducer using a vial filled with a conducting fluid.

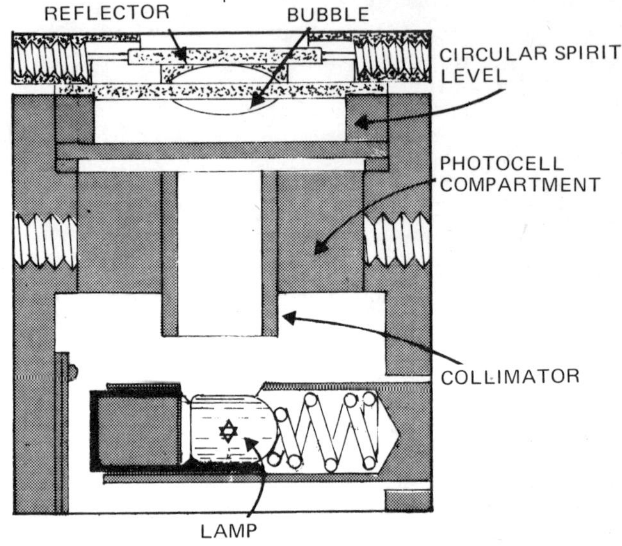

Fig. 5. The N.A.S.A. biaxial tilt transducer. This is capable of 15 arc seconds resolution in each of the axes of tilt.

detectors, monitor the reflected light distribution providing readout of angle in both directions (Fig. 5).

If the depth of the liquid is great, such as in the sea, changes in height can be monitored indirectly using the change of pressure head above a point deep down. In oceanographic research the amplitudes of tides and swell are measured this way using a recoverable capsule which is placed on the ocean floor. In this is a pressure gauge transducer, recorder and power supply.

PENDULUM REFERENCES

As mentioned earlier, the vertical direction is directly related to the horizontal so pendulum devices can be used to measure tilt of the horizontal. Many tiltmeters make use of pendulums. The most straight forward type of pendulum is a mass hanging on a light suspension. Microdisplacement transducers are used to determine the position of the pendulum relative to the mounting frame. If size is not important, the pendulum can be as much as a metre or more in length to increase the sensitivity. Several tiltmeters are available commercially (at a price of many thousands of dollars) that can be lowered into a vertical borehole. In this application, liquid level devices would not be suitable due to the limited size of hole available. It is most important that the pivot point of the pendulum is precisely defined, for the angle of tilt is inferred from the displacement at the lower end together with length of the pendulum. The forces needed to deflect a pendulum are extremely small. For this reason feedback measurement is often used in which the pendulum position is restored by electromagnetic means in a force-balance technique. This helps to ensure that the measuring transducers are always in the same force-exerting position. Borehole tiltmeters often measure the tilt of two perpendicular directions.

If the microdisplacement device is extremely precise, as is possible with well developed capacitance micrometer arrangements, the pendulum can be shortened.

Inductive sensing of the pendulum position has been used in the Talyvel engineering tiltmeter. In Figure 6, a pendulum is used astride an alignment telescope to define a level line of sight.

A plumb-line, as well as defining a line in the vertical direction, is a tiltmeter pendulum of relatively

Fig. 6. An inductively-sensed pendulum tiltmeter being used with an alignment telescope to provide a horizontal line of sight.

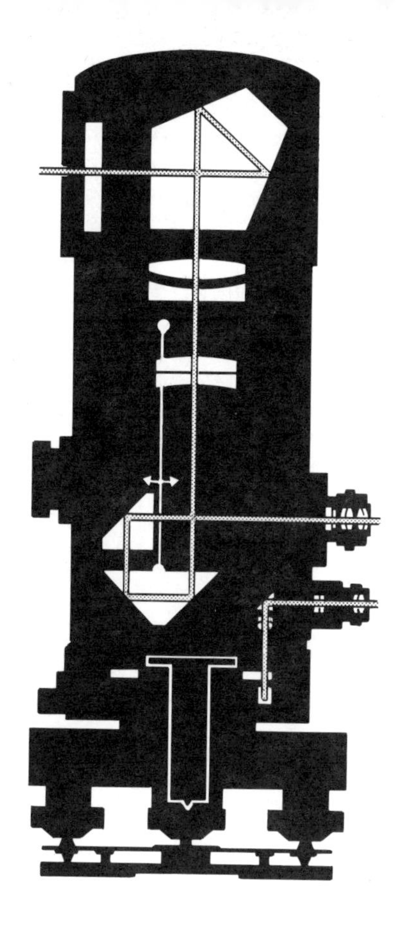

Fig. 7. Schematic of an early Zeiss (Jena) automatic level. The prism always hangs vertically ensuring a level line of sight through the optical system. No electronics is used.

that the wire forms the moving armature of a differential reluctance displacement transducer. They reported a centring accuracy of around 10 μm which is the tolerance limit required in high precision engineering.

It is possible to build mechanical gain into a tiltmeter so that a larger movement occurs in the output member than in the member being driven by the tilt change. Before electronics, these methods were in vogue as there were no other ways to obtain adequate amplification of the small movements. Nowadays, however, electronic displacement transducers can easily sense the fine displacements resulting.

In surveying, the traditional bubble levelling instrument is being replaced by self-aligning or automatic levels. These use optical prisms which are suspended with fine wires so that the optical path always looks out in a level line to a precision of up to a few seconds of arc (in the precision models). There are a number of different methods used, Figure 7 shows just one. Surveying instruments are good examples of how other than electronic solutions to measuring problems may be the better to employ. In this instance the overall weight and, most important, the cost is less than the equivalent electronic method.

PLUMMETS

Before going on to alignment devices, a brief description of the devices used to define vertical sight lines is needed. We have already encountered the automated plumb line. Other methods use optical techniques to define a line perpendicular to the horizontal. In astronomy the vertical is defined by using the surface of a large mercury pool as the horizontal reference. Similar, but much smaller pools are

greater length. Several instances of automated plumb bobs exist. The Russians have published details of a highly precise plumbing arrangement used during the erection of one of their large nuclear accelerators. It consisted of a steel wire plumb line having its bottom weight immersed in a damping fluid. A little above the bottom are two C cores arranged so

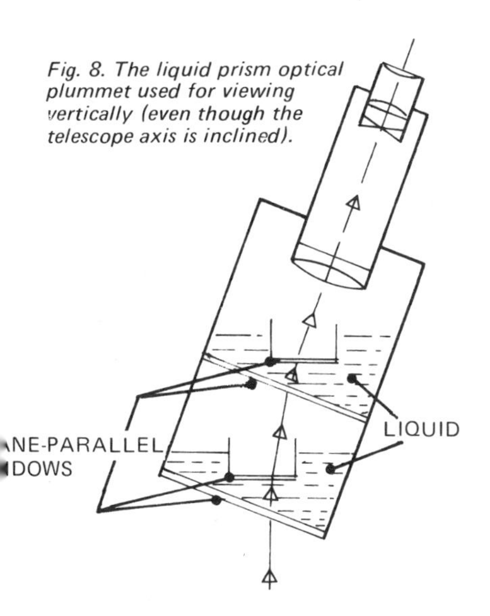

Fig. 8. The liquid prism optical plummet used for viewing vertically (even though the telescope axis is inclined).

Fig. 9. The Dynalens image compensator removes vibration problems when using high powered telescopes.

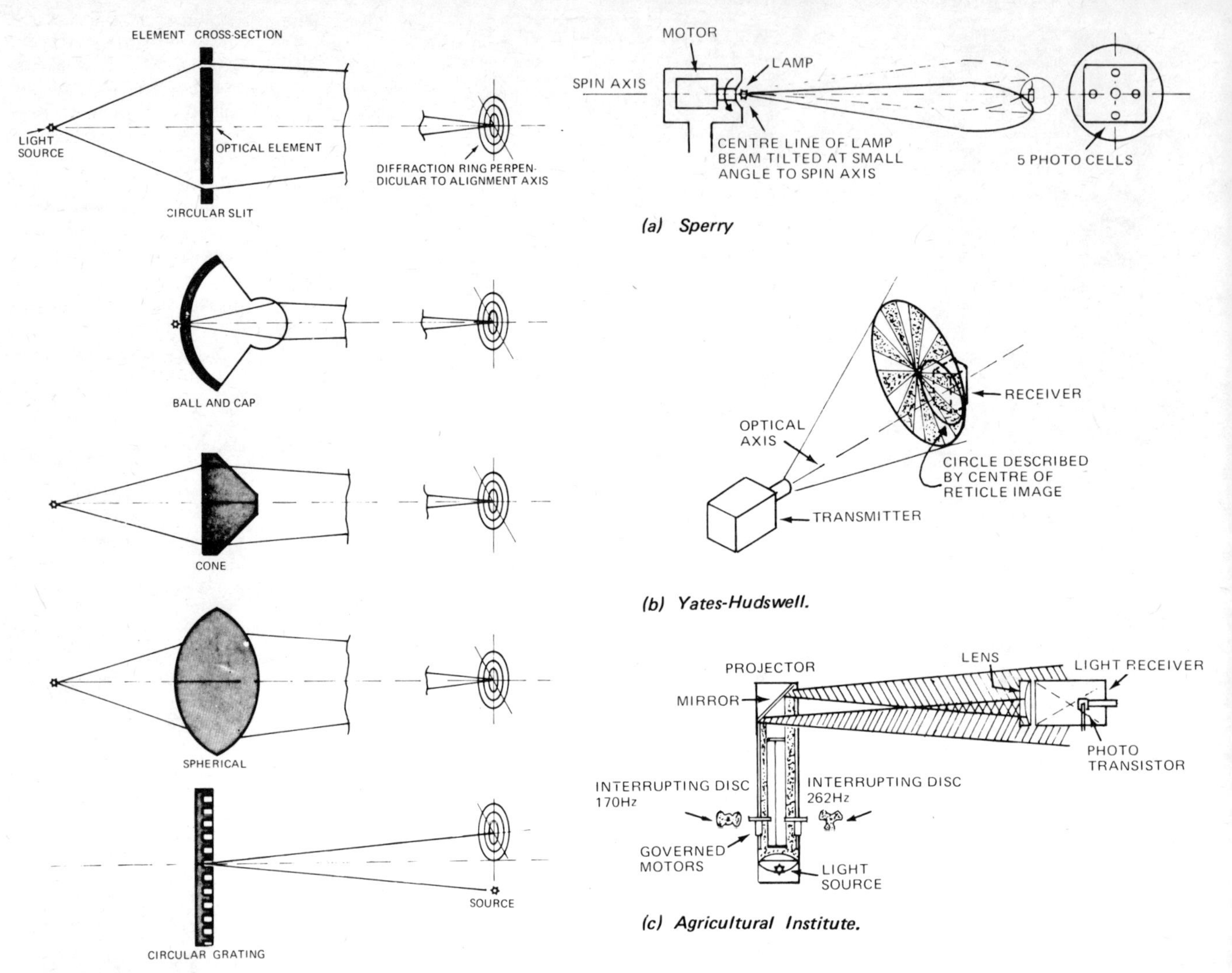

(a) Sperry

(b) Yates-Hudswell.

(c) Agricultural Institute.

Fig. 10. Optical interference alignment devices shown schematically.

Fig. 11. Spatially modulated optical alignment methods using conventional light sources.

made for use with telescopes in the engineering optical-tooling setting-out procedure. At the top of the automatic level shown in Figure 7 is a penta-prism which provides a right angular optical path by careful manufacture of the angles between the surfaces. If the penta-prism is omitted, the visual path will be in the vertical direction. The same instrument is, therefore, easily adapted as an automatic optical plummet.

In mountain surveying operations it is often necessary to hover a helicopter above a ground mark (with high precision) so that an electromagnetic distance measuring instrument can be used to read distances between ground stations. This is not easy, for the helicopter can wander in all six degrees of freedom. One solution is to use a ground-based television camera that is directed in the true vertical direction. A television monitor in the helicopter enables the pilot to see that the ground based camera views a reference mark on the underside of the helicopter. He hovers to keep the mark central in his screen.

IMAGE STABILISATION

An optical device which is useful when the observational platform is unstable is the liquid prism plummet. Figure 8 shows its principle. As the telescope tilts, the liquids flow to form changing dimension prisms which diffract the sight path in proportion to the tilt of the system. This helps to reduce the vertical sight path error caused by the tilt of the platform. Floating windows are used to overcome the vibrational effects on the surfaces of the liquid.

Gyros can be used to stabilise the position of a mirror, and in cases where the vibration is at a high frequency, a mirror spinning in the plane of its surface acts as a good reference. This method is used in bombsight equipments. It has also been used in high power binoculars to stabilise the sight paths. The gyroscope and the liquid lens have been combined in the Dynasciences Corp., Dynalens image motion compensator. Inbuilt inertially-stabilized directional references are used to sense deviations from the steady state position. Error signals actuate liquid prisms, via electromechanical means, changing the optical axis to keep the sight path constant in space. This method has a much wider response bandwidth enabling it to cope in a broader range of circumstances. The dramatic difference between stabilized and

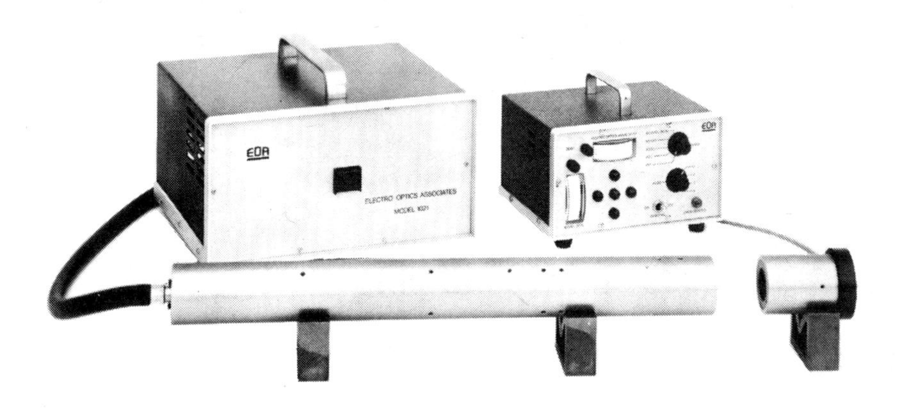

Fig. 12. Two axis, electrical readout, optical tooling laser.

unstabilized scenes, as viewed from a helicopter, is evident in Figure 9.

ALIGNMENT

A large number of situations require knowledge about straightness, flatness, parallelism, levelness, roll, pitch or yaw. At the high precision end of capability is the need to measure the flatness of machine tool beds, surface plates and optical components. Other applications such as agricultural drainage, pipe laying, pipe and pile borer guidance, road grading, concrete slipform paving and railway track tamping also share similar basic needs. Each require methods for measuring deviations from a given line or plane for measurement or automatic control purposes. Such techniques are called alignment methods and as always, no single method suits all cases, so various methods have been developed.

WIRE GUIDANCE

A tightly stretched wire provides an accurate line when viewed in the vertical plane. Special microscopes are available to measure deviations from the wire. These use a double image system that is correctly positioned when both are coincident. In the horizontal plane, allowance must be made for the catenary sag of the wire. The Sulzer factory in Switzerland issues a chart for the sag values. British and Russian reports claim alignment precisions of 20 μm along distances of 50 m by this simple method.

A number of concrete paving and kerbing machines use preset wires to define the road level. Microswitches, actuated by electromagnetic wire position sensors, control the raising and lowering of the paving slip edge. The precision is not high but the requirements for such cases need only centimetre control. As the speed of road and rail transport rises, closer control is needed. In the high speed experimental British Rail track, millimetre precision has been achieved using slip-form pavers.

OPTICAL TELESCOPES AND COLLIMATORS

A graticule placed in the viewing system of a reasonably powerful telescope provides the observer with a line of sight. Precision telescopes made for alignment (one is illustrated in Figure 6) have the optical axis precisely located with the axis of the body. Inbuilt graticules and optical micrometers enable the observer to view targets placed along the line, for example, the bearing housings of a large engine.

An autocollimator is similar except that it has an inbuilt light source that radiates an image of a cross hair onto a mirror (placed at the point of interest) where it is reflected to appear in the viewing field alongside the original image. If the two images are coincident, the outward and return beams are in the same line of sight and the reflecting surface must be square to it. The autocollimator is, therefore, a sensitive angular measuring device. One of these devices is shown in Chapter 3 – where it is used for setting the position of an angle encoder using a reflecting polygon. By using reflectors that remain at the same angle to their base, it is possible to measure flatness or straightness by calculation from the angular tilt of the surface (at chosen locations) and the position on the surface. Automatic readout autocollimators are available that have built-in position-sensitive photocells to give an electrical signal corresponding to the angular deviations of the reflected beam. A specially built unit, operating on this principle, is used to align spacecraft to within an arc second over a range of several hundred metres.

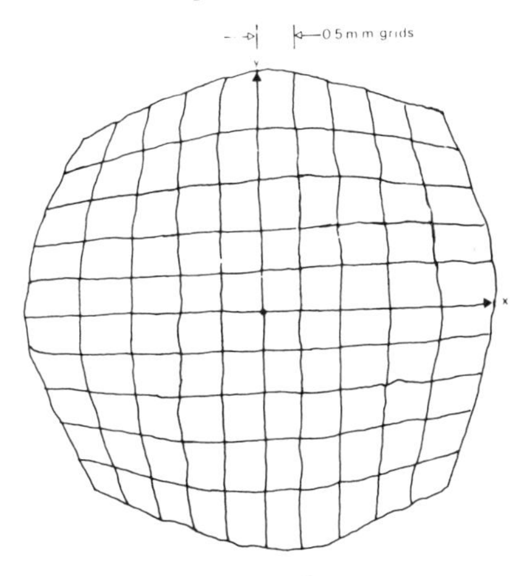

Fig. 13. A linearity plot of a two axis laser alignment readout system.

OPTICAL INTERFERENCE METHODS

In 1950, it was suggested in Holland that a circular slit would produce a diffraction ring along the optical axis, and that could be used for alignment, (Figure 10). The method works reasonably well, enabling alignment to be checked by viewing the fringes with a specially marked piece of perspex. The major defect was that many rings are formed and so the light energy in the central, important one of interest, is not large. By using a number of slits made as close as possible to the shape of the diffraction rings produced when light shines through a pin hole, it is possible to concentrate much more energy into the central bright spot or ring. These plates are known as zone plates. The National Physical Laboratory in Britain use these to provide a reference base, several hundred metres long, upon which alignment devices may be tested. There are many other ways to produce such rings and a few are shown in Figure 10. The latter is marketed for industrial use. Current research aims at providing electrical readout from such rings by scanning across them in order to determine the best centre.

Spatially Modulated Systems

If a beam of conventional light is radiated, it diverges to such an extent that the centre is difficult to detect with precision. Furthermore, the amplitude will vary with time due to changing atmospheric conditions. For this reason, various methods have been evolved in which the beam is modulated in a spatial manner.

A system developed by Sperry is shown in Figure 11. A motor rotates a slightly tilted optical system so that the optical axis nutates in space. A detector having five photocells is placed on the beam axis. If central on the detector, each of the four outer

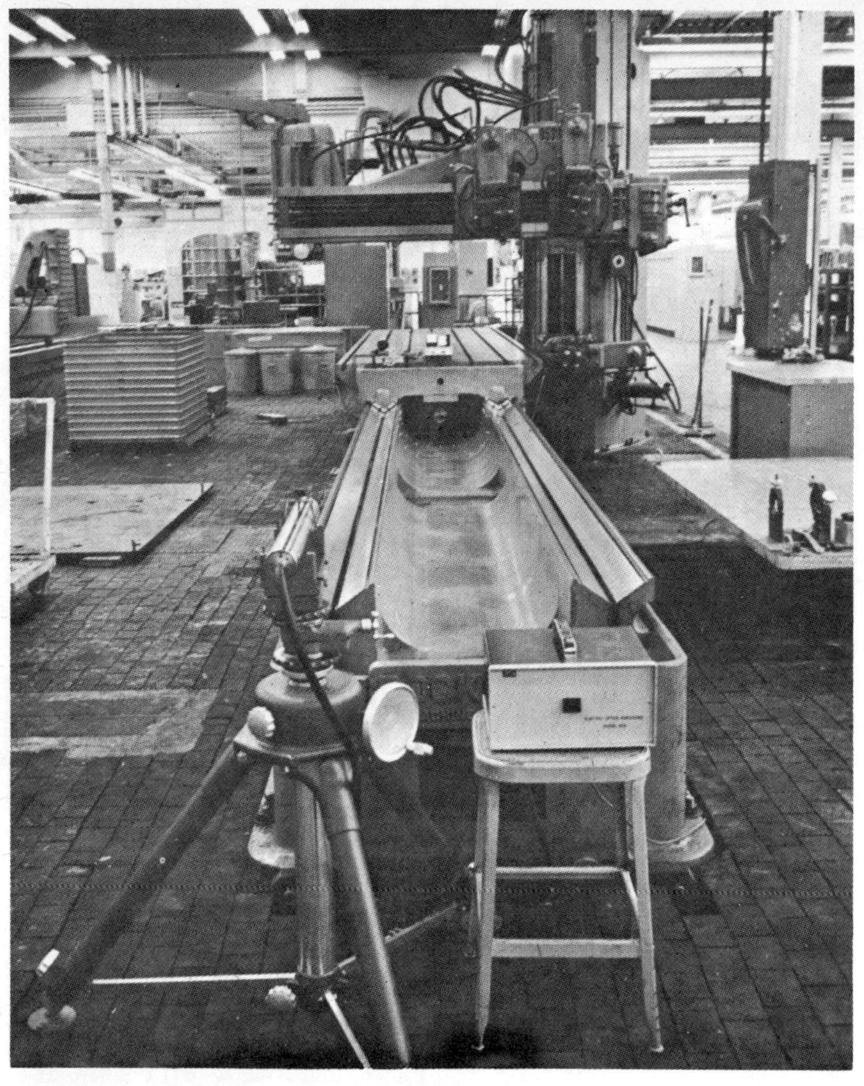

Fig. 14. Testing the straightness of travel of a planer.

cells has an equal ac component and mark/space ratio. If not aligned, error signals are produced. This method can resolve to 0.3mm in 30m distances and has been applied to an experimental road grader to hold the blade in a straight plane regardless of road surface. A pendulum sensor holds the blade's transverse angle in control.

The next method is adapted from a military tracking system. It has been used to control the ploughing-in of agricultural pipes in Britain. A spinning disk produces a spatial chopped beam that also nutates. This is sensed to see if there is any out of balance ac signals in a somewhat similar manner to the one described above, the difference being that frequency modulation results (not amplitude). It is claimed that this has a better signal to noise ratio in practice, for optical transmission in the normal air environment is less noisy to frequency effects than to amplitude.

A third system, developed by the National Agricultural Research Institute is shown also in Figure 11. In this a light source provides two beams closely parallel to each other. One is chopped mechanically at 170 Hz, the other at 262 Hz. The detector sees a spatial overlap of each frequency. When central it gives even amplitude signals at the two frequencies. This was used in an automatic pipe laying tractor.

Laser Alignment

The automated techniques mentioned above were devised before the laser was developed (remember the laser is only an invention of the last decade) and the sophisticated methods were needed to overcome the lack of an intense narrow beam of light.

Continuous-wave lasers are now used extensively in alignment, for the beam can be collimated to keep the beam width down to millimetres over hundreds of metre distances. For crude alignment, the beam can be viewed by eye (but not directly for fear of eye damage). Electrical output is generally provided by a quadrant silicon photocell position detector. One set of equipment is illustrated in Figure 12. Alignment lasers are now especially designed to ensure that the beam leaves the precision ground barrel in the centre and does not have significant angular variations with time. When using a quadrant detector, the two output signals are interrelated to an increasing extent as the spot moves from the centre. Figure 13 shows a typical calibration plot of the linearity obtained from the x and y axis signals. In the centre the curves are the closest to a square, which is the ideal. A typical set-up using laser alignment equipment is shown in Figure 14. If the time axis of a pen recorder is synchronized with the machine slide rate, the two axis alignment errors can be plotted automatically as the slide advances along the planer bed. Traditional manual methods could give similar accuracy but were tedious and extremely slow.

Alignment methods provide means to measure straightness. If constant grade is desired, for instance, then the alignment device is used in conjunction with a level defining method to set the line in the correct direction and plane.

Finally, on the subject of alignment measurement, mention should be made of the use of inertial guidance. A gyroscope, once set going, produces torques if its non-spinning axes are rotated. By sensing the torques it is possible to hold a straight line in space without the need for a physical reference position. One interesting problem described in the Russian literature was how to monitor the alignment and grade of pipes already buried. This was solved by making a small wheel-driven mole that drove itself through the pipe. It carried a gyro unit, recorder and distance meter. By synchronizing the recorder chart speed with the driving wheel and plotting the instantaneous grade at all times from the gyro, a plot of straightness was obtained.

FURTHER READING
(see chapters 1, 2, 3 and 5)

CHAPTER 5

USING TRANSDUCERS TO MEASURE AND CONTROL POSITION

Inspection machines, such as this unit by Schiess, are used to make multi-axial positional measurements.

SO FAR we have seen how lengths, angles, tilts and alignments are converted into a convenient electrical form for purposes of measurement or control. This Chapter describes how these one dimensional techniques are combined to yield positional information of an object or a point lying in a plane or within a space – for there are many alternatives available where multiaxial measurements are needed. Examples are the shape of an aircraft frame, the positioning of the turbine blades inside a jet engine or a steam turbine, the control of numerically controlled machine tools to form complex shapes, the cutting of steel plates and their assembly into super-tanker ships, the measurement of the profile of radio telescope dishes, control of automatic tractors in ploughing – the list appears limitless.

CO-ORDINATE SYSTEMS

Before describing just how transducers are combined, it is necessary to become familiar with the ways in which a point in space can be defined relative to some reference system by a number of individual, single-dimension, measuring devices.

Position of a point in space can be defined by three parameters using lengths and angles. (A point has no size so orientation is of no consequence.) At least one measurement must be a length to define the physical position. Most important is that three measurements must be made relative to some kind of established reference system. Figures 1a and 1b show the two most commonly used methods using lengths only. The cartesian, or rectangular, system defines the position of P by extending lines perpendicular from each axis to give the x, y and z values. The triangular concept defines P by the lengths S_1, S_2 and S_3 which extend from the corners of a fixed-size reference base triangle. It is also quite reasonable to determine P using the angles between the base triangle and the S sides if the length of the sides of the base triangle are known. The concepts of rectangulation and triangulation are often combined for reasons of expediency. For example, using the cartesian framework we could have an $\mathcal{R},\Theta,\phi$ polar system (Figure 1c) in which the length to the origin is $\mathcal{R}$ and the other co-ordinates the angles made between that line and the cartesian axes.

Position in a plane needs only two dimensions, for the third is held constant by definition. Again, at least one dimension must be a length. If the point moves along a line, two of the possible three dimensions are constant, so only one needs considering and this must be a length. Alignment could be considered as a zero-dimensional measurement, for no lengths are measured along the line. It is, however, really a 2-D case as deviations in a plane perpendicular to the line are the parameters of interest.

If the object of interest has physical size, the orientation of its shape in space is also important. As three

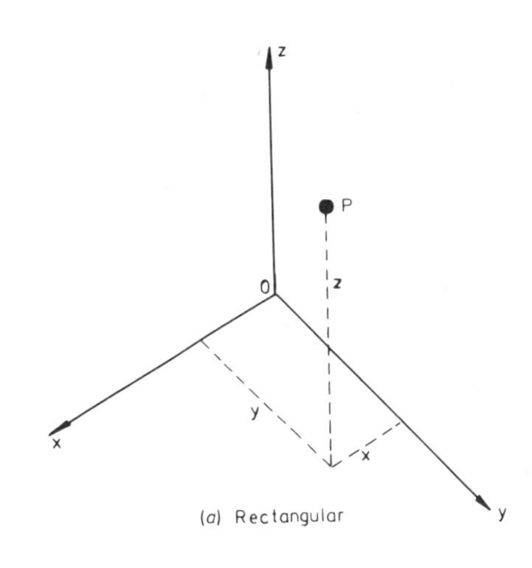

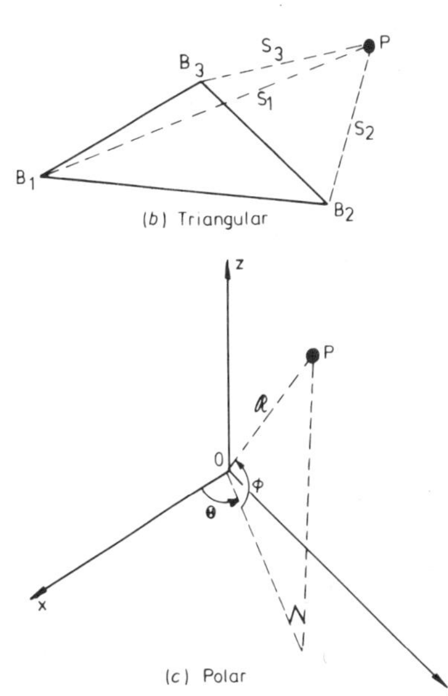

Fig. 1. Reference frameworks for multiaxial positional systems.

Fig. 2. A Bendix "Cordax" measuring machine being used to inspect the critical dimensions of a machined part.

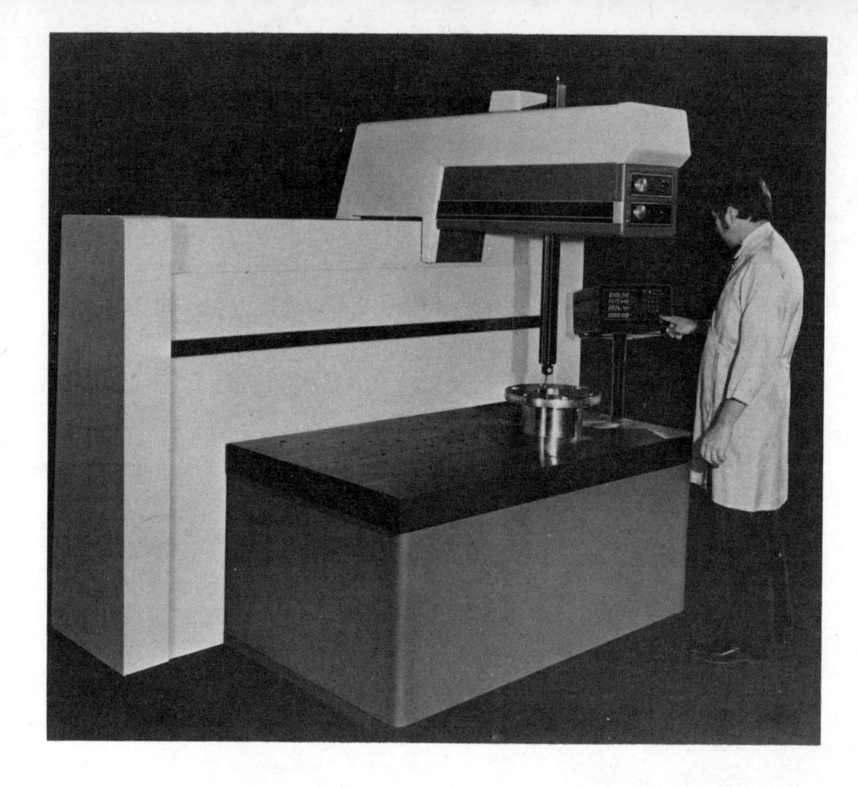

degrees of rotation are possible, it may be necessary to measure as many as six variables to define position adequately. A missile in flight will have pitch, roll and yaw components, as well as dimensional position values. By now, it should be clear why lengths and angles are inter-related in the practical measurement of position. The choice of reference used is largely a matter of convenience. Rectangular systems become difficult to use when sizes extend beyond several metres. Above this, polar and triangular systems come into their own. Smaller range measurements are able to utilize the rectangular arrangement for it is economically viable to manufacture the necessary mechanical reference framework.

RECTANGULAR METHODS

The bulk of industrial machines built for making or measuring work-pieces use rectangular co-ordinates. In Chapter 2 a large numerically-controlled machine tool was illustrated. Smaller units are more usual. A precision inspection machine having three-axis digital readout and recording facilities is shown in Figure 2. High grade machines such as this are accurate to a few micrometres when used correctly and given the right environment. Translating axes are provided as a rectangular framework having separate length transducers on each axis of movement.

The end of the stylus is the position in space being measured, and for this to be accurate, the three slides must be straight to within a micrometre, and square to within seconds of arc. Furthermore, the travelling cantilever arm must hold this angle as it translates: yet it must be stiff enough not to sag significantly as the vertical axis is moved out along its arm. These requirements can be met but at considerable cost. Inspection machines need to be more accurate than the manufacturing machine that produces the components to be checked, but as there are no machining forces involved in the structure, they can be lighter in construction. Inertia of the slides is kept to a minimum to allow the operator to move the stylus more rapidly. The framework of a numerically-controlled tool, however, needs to be especially stiff, for the inertial forces produced by its rapid movements greatly exceed the static forces. Dynamic accuracy is important to retain precision and stability from the control system.

The requirement however, is not always for three axis measurement. In printed circuit-board inspection or drilling, in map making, in bubble chamber photograph digitizing and in automatic flame cutting of sheet, to name just a few examples, the need is for only two axis measurement.

Automatic draughting machines are used extensively and many companies are in this market. Coupled to a computer they are able to produce drawings of extreme precision and complexity. For example, integrated circuit manufacture relies on them. An example of precision artwork of a Honeywell integrated circuit is shown in Figure 3. Another use for drawing machines is to check out numerical control tapes before they are used on the machine tool. This is especially useful in the ship building industry where individual varied shapes are nested together on a single stock-size plate.

If the two axes having readout are free to move, rather than being held by position servos, the machine can be

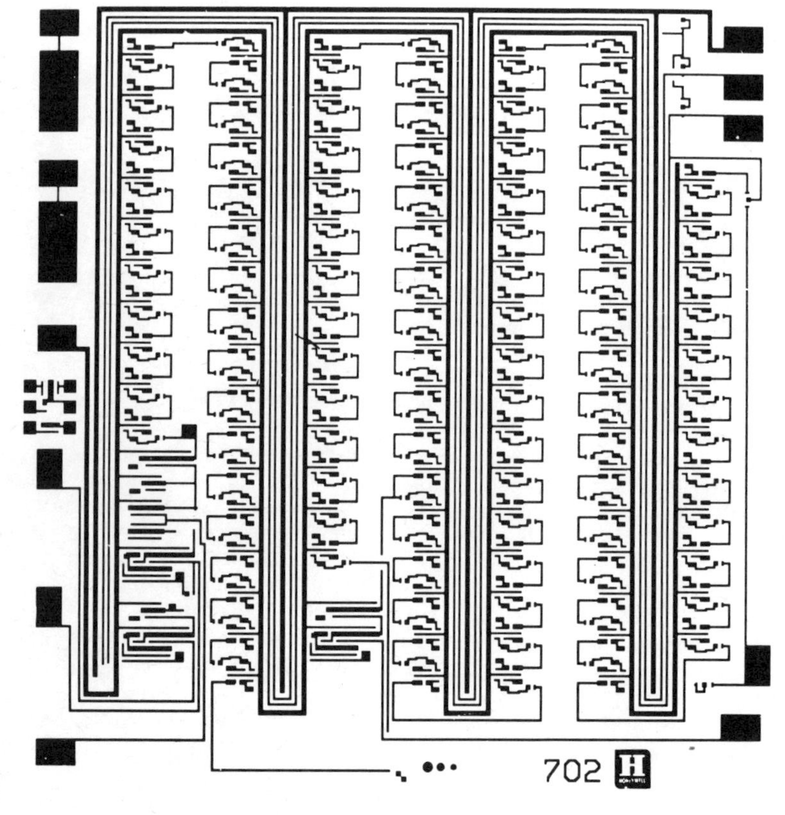

Fig. 3. Precision artwork generated on a Gerber automatic draughting machine.

used to digitize drawings. Several drawing readers and chart digitizers are available. One unit is pictured in Figure 4.

It is also possible to build a three dimensional equivalent of the drawing reader that can produce co-ordinate data as well as drawings of a free-form shape. Several car factories have automatic tracing units, one type is illustrated in Figure 5. The part to be drawn (a clay model or plaster flash of a panel) is placed inside the framework and the stylus driven down into contact. Whilst the horizontal carriage is being driven along over the part, the vertical axis follows the surface contours automatically. On each axis are length transducers that provide the three coordinate values on punched tape. An unusual feature of the unit shown in Fig. 5. is that it also draws the three views as it moves; most machines draw them on another plotter. On each side of the frame is a drawing table. The common axis of the two side view drawings drawn on each board is mechanically tied to the stylus, the other axes being electrically linked to the cross horizontal and vertical movements. The third drawn view is produced on a separate end-view board controlled by the two electrical signals.

IN-SITU OR COMPONENT-SUPPORTED MEASUREMENT

As the size of the part to be measured or manufactured increases, it becomes increasingly expensive to hold the framework stable and accurate. For instance, 20 metre bed boring mills are made at the rate of three or four per year but their cost of a million dollars each is rarely justifiable. Secondly, not all multi-dimensional objects can be moved to a machine for measurement or manufacture. The surveyor cannot take a building plot into the office to measure it. So for large sizes the object is measured by taking the measuring devices to them, mounting them upon or around them.

In large-scale industrial measurement this has become known as the component-supported or in-situ method of measurement. The work piece acts as a stable precise bed frame and only a small work head and measuring system is needed to provide measuring and precision manipulation facilities.

In the optical tooling procedure, light but accurate slide ways with calibrated scales are placed around, say, an aircraft frame to form x and y axes. On these move precision alignment telescopes, (Chapter 4) that project a point on the fuselage out to each scale. In this manner the contour of the airframe can be checked.

This procedure has been automated by the British Oxygen Company for flame-cutting large plates by computer control. Their experimental equipment is shown in Figure 6. Two precision slides carrying motor-driven carriages are mounted at right angles around the plate to be cut. On the top of each is a telescope having an inbuilt optical position-sensitive detector that senses small errors of position between the carriage and the vertical strip light source seen on top of the cutting head. If the light source is not exactly at the intersection point of the lines of sight of the two telescopes, error signals are generated that redirect the tractor. In this way the cutting head is made to contour by following the carriages. The tractor itself is worthy of description. On each side of the square support frame are driving tracks. In the track links are small rollers that enable the track to slide sideways whilst driving forward. In this arrangement the direction moved by the tractor depends upon the relative velocities of each pair of tracks. The overall concept is capable of control over very large areas, the limits being set by the optical turbulence of the sight paths. Alternative methods of performing the same task have been developed as will be seen below.

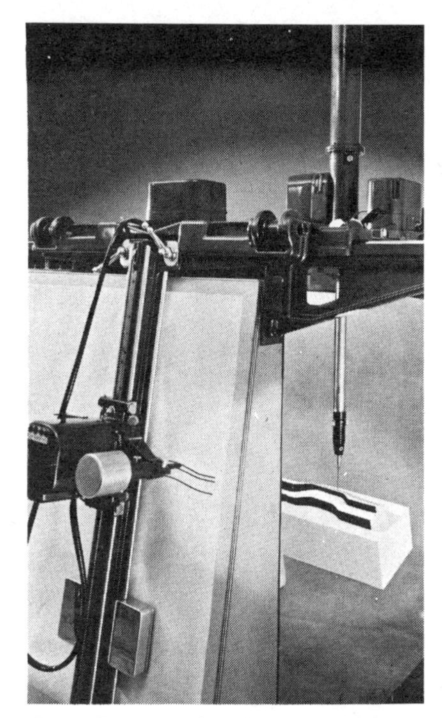

Fig. 5. This type of tracing and digitizing system is used in many automotive plants.

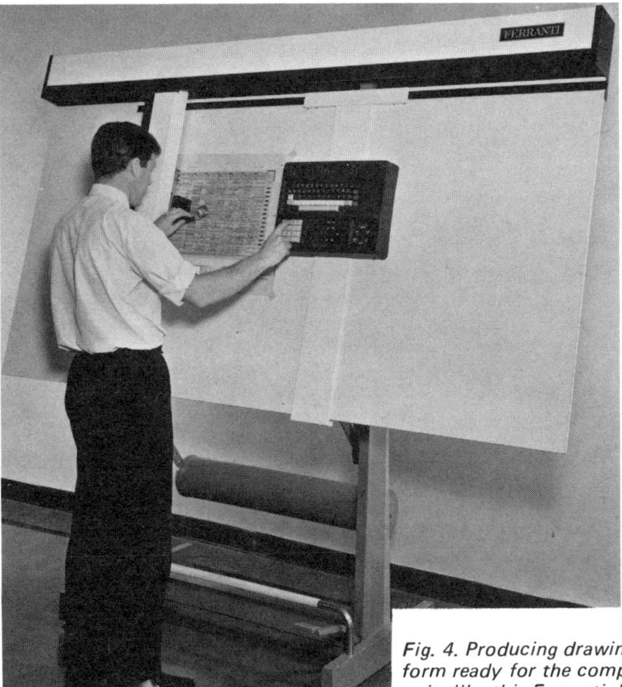

Fig. 4. Producing drawing coordinate values in a form ready for the computer is achieved with units like this Ferranti drawing reader.

TRIANGULATION METHODS

If position within a plane of large extent is needed, the use of a triangular basis of measurement is attractive, for only two fixed points are needed as a reference instead of two slideways. There are many alternative combinations for defining the points position. If a base line is fixed, two lengths will complete the triangle giving the position of the apex with respect to the ends of the base line. Other combinations use two angles and the fixed baseline or one side and one angle between it and the base line direction. In three-dimensional triangulation there are over twenty different schemes so the choice depends upon other factors of individual applications. For example, a radar unit can most simply track an object in flight by using the $\mathcal{R}$, Θ, ϕ arrangement shown in Figure 1, for only one radar unit is needed. Alternatively, two units at the ends of a known length baseline are often used

Fig. 6. Automated optical-tooling procedure for cutting large steel plates by computer control.

Fig. 7. A polar co-ordinate digitizer needing no framework.

to avoid the need to measure range distance.

One important factor that needs consideration is the attainable precision of angle measurements versus length measurements in such cases, for extreme precision angle transducers are costly. Let us now consider equipment that has proven practicable using non-rectangular methods.

Coradi, a Swiss company who specialise in mathematical instruments, have marketed a polar co-ordinate drawing digitizer that has the convenience of being usable merely by placing the unit on a drawing as shown in Figure 7. The operator places the cursor cross-hairs over the point of interest, position being recorded as the length of arm extended from the frame and the angle it makes with the base plate.

In nuclear research, bubble chambers are used to record tracks of nuclear reactions. These vapour trails are photographically recorded and their positions digitized ready for processing in digital computers. Each exposure needs hundreds of points to define it and literally millions of photographs are taken each year. Data processing is, therefore, a major problem in this research. One solution to the digitizing problem which yields a moderate rate of information coding uses a device first developed by a French team who called it a "Bidule a fil." Later it became known by the Italians as a "Mangiaspago" or "string-eater".

A schematic of one of these devices is shown in Figure 8. Thin wires, tensioned with hanging weights, pass around drums placed at the ends of a base line member. The two wires join at a viewing puck to complete a variable size triangle. As the puck is moved, each wire rotates a drum which is connected to an angle transducer thus giving a signal equivalent to the length of the wire between the puck and the drum. The Brookhaven Laboratory units are improved versions of the earlier French equipments and can digitize position within an area of 150 by 60cm – to within 10 μm.

The potential of triangular measurement in industrial automatic control for large component inspection and manufacture was recognised independently in the late 1960's. Figure 9 shows an experimental trilateral equipment that can control its tool head position (seen on the cantilever arm of the printed-armature motor-driven tooling actuator) to within a few parts of a million in areas of up to 20m by 10m. The demonstration angle-iron base line holds two fast-response spring-tensioned wire-drum length transducers at a fixed distance. The wires join to form the apex of a triangle by connection to two large-bore ball races. These effectively project the wires to form the triangle apex while allowing a tool to be placed at the intersection, thus observing Abbes principle of direct measurement as closely as possible. On the right is the electronic unit that derives two-axis cartesian coordinate error signals from the trilateral wire error signals. These are then used to control the axes movements of the portable tooling head.

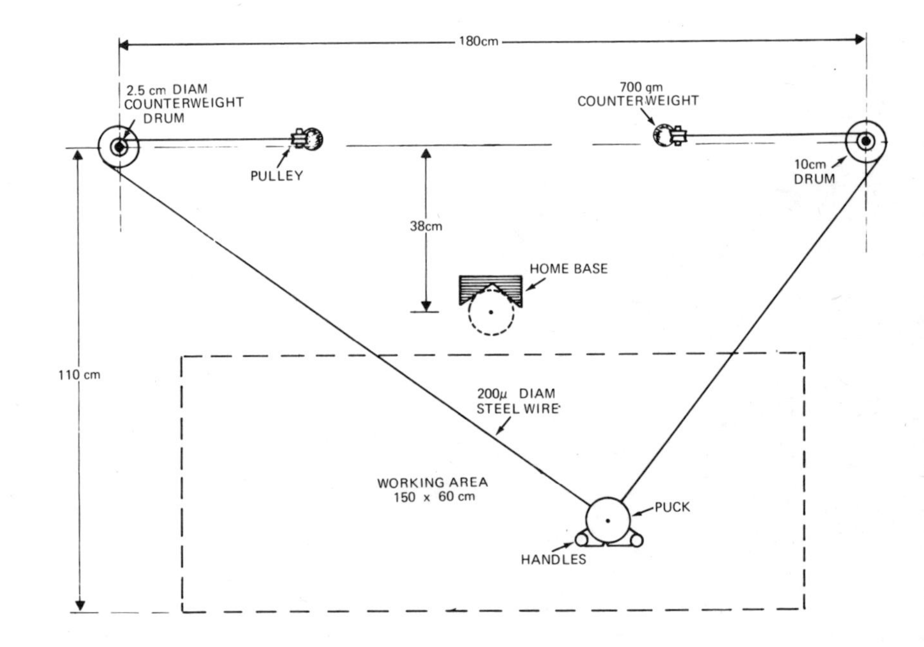

Fig. 8. Schematic of the "Super Mangiaspago" of the Brookhaven National Laboratory.

Fig. 9. Automatic position control system using trilateral co-ordinates.

The shape of some large engineering structures dictates a need for non-cartesian methods due to sheer size and immobility. Radio telescopes are good examples of cases needing extreme accuracy shape measurements. The Arecibo telescope in Puerto Rico is over 300 m in diameter, and proposals were recently under study for resurfacing it to an accuracy of 3 mm. Smaller, but still large, units such as the steerable 64 m Parkes telescope in Australia have a surface accuracy of 5 mm. The accuracy of the shape largely decides the upper frequency of operation and the antenna gain, so it is most important. Absolute shape is initially obtained using conventional surveying methods, but special devices have been produced using triangulation formed visually with pentaprisms and ruled scales. Shape changes under various loading conditions were measured in Australia using a central scanning camera that photographed the relative positions of hundreds of reflective targets placed over the dish surface. To date it has not proven economic to build automated measuring systems for the dishes. Photogrammetry has proven useful but this also is not completely automatic. This method is outlined later.

Surveyors rarely use rectangular co-ordinates, for they are inappropriate due to the difficulty of defining the x, y, z axes lines from which a line is projected to the point of interest. Instead tapes and electromagnetic distance (EMD) measuring instruments (discussed in Chapter 2) are used to define distances between triangulated bench marks, in conjunction with angles measured by manual or digital recording theodolites. In geodetic survey, trilateration (lengths only) is now used predominantly as greater precision is possible with E.M.D. devices for a given amount of effort.

Satellites have also been employed for accurate global measurement. Reflecting balloons 41m in diameter, and known as Echo, are spotted on film from stations located around the Earth. Data derived from the time-correlated satellite and star image background enable the position of the Earth stations to be determined relative to each other. An observation station in Canada is shown in Figure 10. These methods are able to measure to within a part in a million over intercontinental distances. The photographic plates are measured with small x-y digital readout co-ordinate tables.

Fig. 10. The Whitehorse satellite tracking station in the Yukon. Precision cameras photograph the relative positions of the satellite and its star background.

But what if observation between ends of a line is not possible, such as inside a mine? In these cases inertial navigation can be used. The gyroscope, shown diagrammatically in Figure 11, provides a fixed direction reference in space. Boeing 747 superjets, Mace and Titan missiles, space vehicles, marine vessels and the military, each use inertial guidance. In the Boeing Carousel IV navigational unit, three accelerometers measure the magnitude of the aircraft acceleration components from which velocity is obtained by integration and position by a second integration. Three gyros hold the accelerometers in the same spatial directions. These units operate with a reliability equal to running a colour T.V. set for forty years without failure. For an hour of flying, the navigational error is less than 2 nautical miles.

Gyros are also used in conjunction with theodolites in surveying situations where triangulation is difficult. The unit shown in Figure 12 has a gyro reference in the underhung cylinder below the theodolite. This unit can define directions to within 1 min of arc for a 15 min observational period. If longer observations are made, it is possible to obtain 10 arc second accuracy.

PHOTOGRAMMETRY AND AUTOMATIC MAP-MAKING

Until recent years, photogrammetry was confined to use in topographical mapping. An aircraft flying at constant speed is used to take aerial photographs of over-lapping areas of ground as it flies. Successive, high-quality photos are then viewed together in a special way as a stereoscopic pair, producing a three-dimensional scene to the observer. Stereoplotting machines convert the pictures to maps. With these machines, the operator controls the movement of a floating point in the 3-D view and moves it over the contours of the 3-D image as though following a path in reality. Photogrammetry is a way to scale down (or up) a 3-D object and reconstitute it somewhere else. The x, y and z co-ordinate values are either recorded or plotted as maps via mechanical or electrical links. The equipment is costly as can be realised from the illustration in Figure 13 — but map making is a big industry.

In the last decade, photogrammetry has been used for other purposes; in medical research for physiological shape change assessment, for checking radio dish and car body shapes and even for recording car accident scenes.

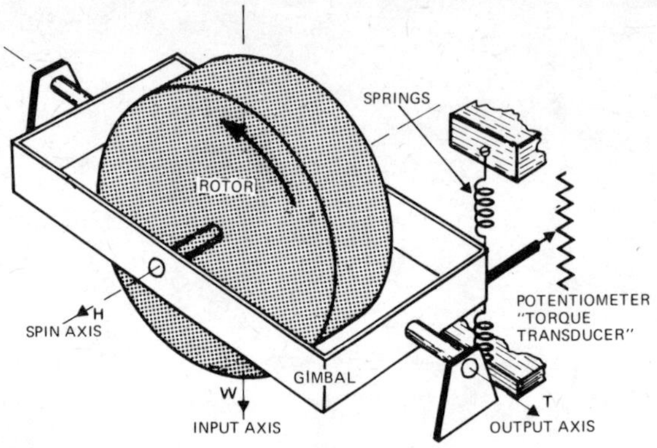

Fig. 11. Schematic illustrating the gyroscope as a reference direction producing device.

Fig. 12. A gyro-assisted theodolite.

In Australia, B.H.P. use it for monitoring the stock piles of ore, etc at the end of each day's operations. In the States it has been used to map the sea floor by lowering cameras to the bottom.

As can be guessed, operating the stereo plotting machine is an exacting task requiring a skilful operator, and in recent years research has been directed at automating the process from photo-pairs to finished maps. The difficulty is that although the human computer is not fast or reliable compared with electronic computers, it does have the power to handle vast numbers of variables and come up with a decision using unknown methods. Machines have difficulty in deciding what to do in mapping tasks, so maps made automatically still have a lot to be desired because of the limited rules that can be programmed in.

HOLOGRAPHY

About twenty-five years ago Gabor proposed a method known now as holography. In this, coherent radiation is used to illuminate an object. Some of the original radiation is optically mixed with that reflected from the object, thus forming a two-dimensional interference pattern which looks nothing like the object. This is recorded on film as a hologram. If the hologram is viewed with rear illumination from a coherent source, the object is apparently reconstructed as a 3-D image having depth and form. In a way the hologram is akin to a stereo pair of photos as they both have recorded a 3-D shape on a plane medium for easier viewing elsewhere. Holography, however, operates on an interference basis using short wavelengths and, therefore, has extreme resolution. For example, in time-resolved holography, an exposure of the object is made. The developed picture is the combined pattern of two holograms and exhibits Moire-fringes representing surface errors of small magnitude.

The method has been used with visible radiation, for testing cylinder liner accuracies, turbine blade stresses, optical mirror blanks, for studying how insecticide falls on insects and for car-tyre inspection.

I.B.M. have a computer programme that produces holograms without an object to start with. These kinoforms, therefore are synthesised visual experiences.

Holography is not confined to the visible. Microwaves and radio frequencies can be used for seeing in the fog or in darkness or for looking into opaque materials. Ultrasonic radiation has been used for mapping the geology and sands of the sea floor in exploration uses.

SCANNING SYSTEMS

At the smaller end of position measurement is the electron microscope which magnifies the size of an object so that it can be more readily observed and measured by the unaided eye. The most recent technique is called scanning electron microscopy for it uses a scanning principle to build up a two dimensional picture of an object with only a single-point intensity determining detector. A block diagram of the Cambridge Stereoscan instrument is given in Figure 14. The electron beam is focused to an ultrafine probe on the specimen. Scanning coils deflect this beam to sweep it in a line across the surface of the object. Emitted (or transmitted) electrons are collected as a current related to the reflectance (or absorption) of the specimen. Successive scan lines are placed side by side on a C.R.O. tube to build up a 2D picture. This concept of scanning is common to conventional television cameras, remote-sensing thermal radiation scanners, ultrasonic prospecting and physiological diagnostic instruments.

Ultrasonic distance transducing techniques usually use the pulse return method so are able to measure at only one place at a time. In the Sonogram instrument used in livestock fat assessment, a curved frame is held over the live animal's body. The measuring head moves automatically along the frame producing various reflection echoes at each position. A camera also slowly tilts in synchronism and the photographic plate is exposed where reflections occur. The composite picture produced shows the position of the junctions of the various layers of fat and muscle enabling the quality of, say, chops to be determined without the need for slaughtering.

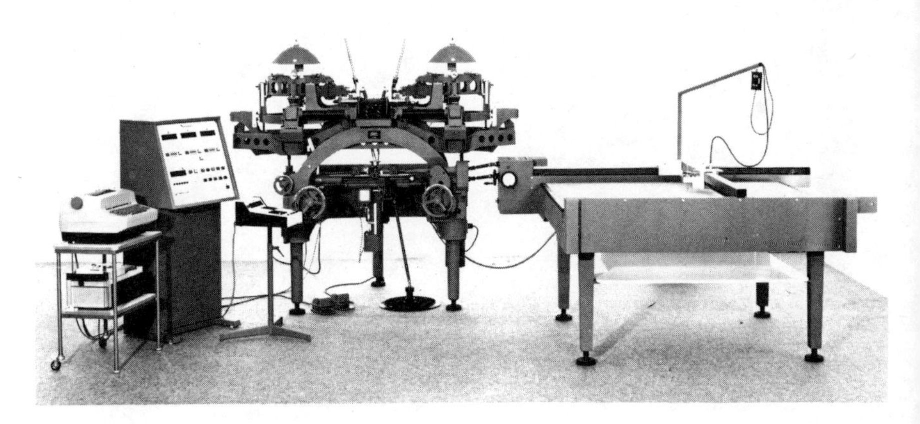

Fig. 13. This machine complex is used to plot maps and other contours from a stereo pair of photos.

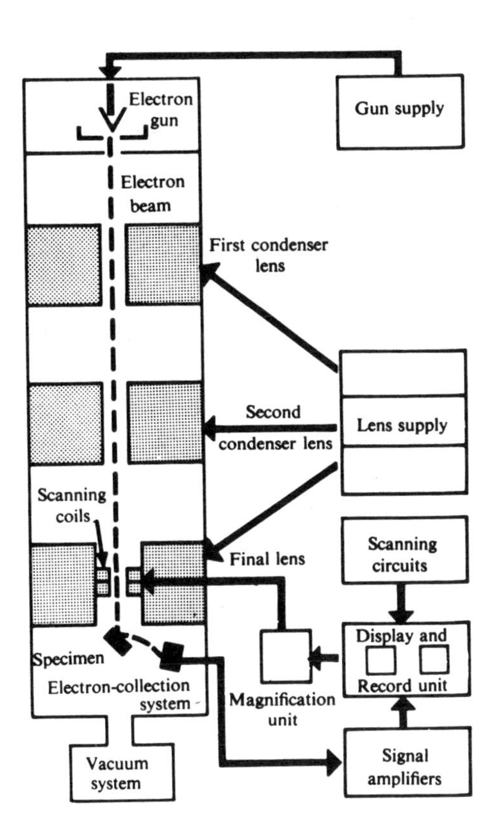

Fig. 14. Block diagram of Cambridge Stereoscan electron microscope.

Similar instruments are used in medical diagnosis for they give different information to the X-ray picture which yields a contrast proportionate to absorption rather than distance information. Although the end result of scanning is a cartesian picture construction, the physical conversion is formed by using a combination of lengths and angles.

FURTHER READING

"Scanning Electron Microscopy : Systems and Applications 1973". W. C. Nixon, Adam Hilger, London, 1973.

"Developments in Electron Microscopy and Analysis 1977". D. L. Misell, Adam Hilger, London, 1977.

"Optical Tooling for Precise Manufacture and Alignment". P. Kissam, McGraw-Hill, New York, 1962.

"Holography and its Technology". J. N. Butters, Peter Peregrinus, London, 1972.

"Metrology with Autocollimators". K. J. Hume, Hilger and Watts, London, 1965.

"Infra-red Systems for Detection, Direction-Finding and Automatic Tracking of Moving Objects". L. Z. Kriksunov and I. F. Usol'tsev, Gordon and Breach, London, 1972.

"The Measurement of Mechanical Parameters in Machines". N. P. Rayevskii, Pergamon, London, 1965.

"Practical Engineering Metrology". Pitman, London, 1970.

"Laser Gauging". P. H. Sydenham, Continuing Education, University of New England, N.S.W., Australia, 1976.

CHAPTER 6

HISTORY AND TECHNIQUE OF TEMPERATURE MEASUREMENT

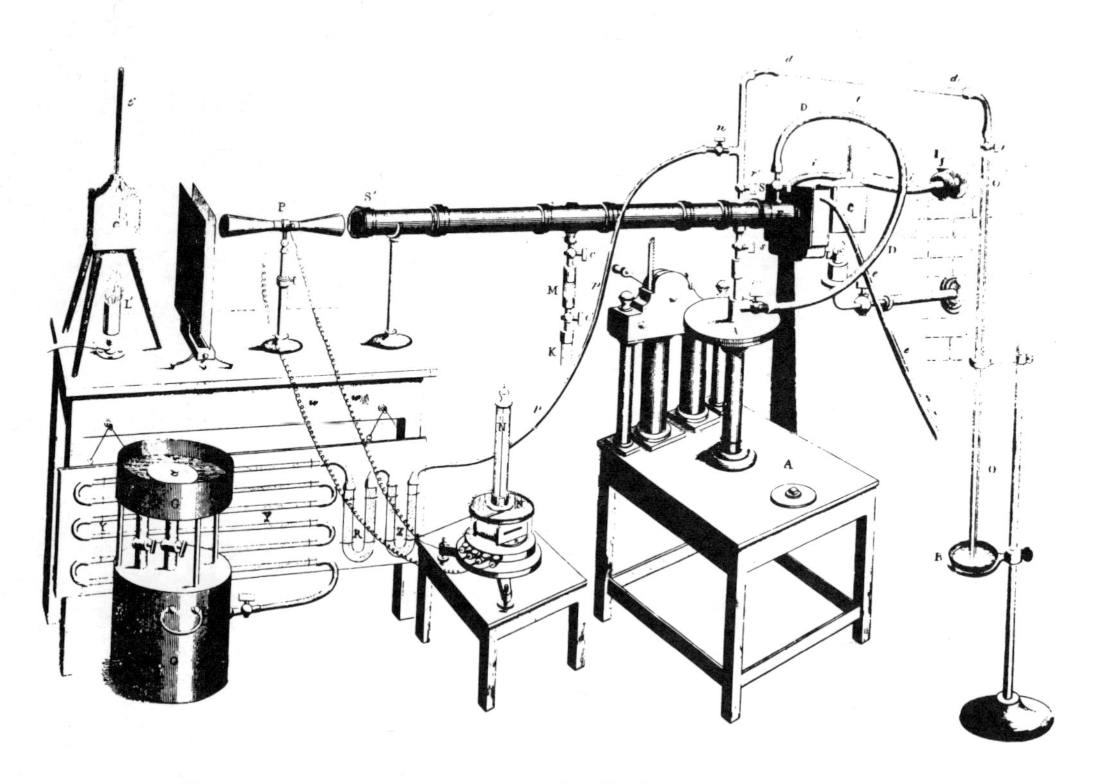

Fig. 1. Just over a century ago apparatus like this of Professor Tyndall was in use for investigating heat phenomena.

PREVIOUSLY we have discussed dimension, which is termed an extensive variable, where two physical lengths (or angles) can be placed together to produce the sum of the two. By contrast there are some physical parameters that cannot be added this way, as addition of equal quantities merely produces the same value. These are known as intensive variables of which temperature is one.

It is now known that heat is a form of energy but it has only been regarded this way since the late 18th century when Benjamin Thompson proposed the energy theory using, as an example, the boring of cannon. Prior to this, a substance known as caloric was thought to permeate all matter: the more caloric present, the hotter the body.

Experimental evidence shows that heat energy is stored as rapid molecular motion. Boyle had suggested this concept in the late 17th century, but it was Joule who finally demonstrated the numerical equivalence of heat energy and mechanical energy in the 1840's.

Temperature is the measure of the degree of hotness or coldness of a substance and a thermometer is a sensitive device used to indicate temperature in a convenient manner. Human senses are very limited when attempting to resolve temperature differences.

When two substances at different temperatures are placed in physical contact, heat flows from the hotter to the colder until thermal equilibrium is reached, that is, they reach the same temperature. This axiomatic law of thermodynamics enables auxiliary devices to be used to sense temperature. In many cases the presence of the thermometer may alter the measurement, so care is needed when using temperature sensors.

SCALES AND UNITS OF TEMPERATURE

As temperature is directly related to molecular kinetic energy, it is theoretically feasible to define temperature in terms of the mass and velocity of the molecules but this is not possible practically as yet. The high precision of mass, length and time standards (errors less than parts in 10^9 in each case) cannot be realised when measuring masses as small as molecules. Other methods of standardizing temperature have therefore been devised.

Practical thermometers existed well before the energy concept was accepted and ancient references credit Galileo with inventing the first around 1590. These early thermometers used the fluid (liquid or gas) expansion effect and it was not until the 1820's, when the basic laws of electricity were discovered, that electrical methods began to appear for measuring temperature. Figure 1 shows how a laboratory was equipped in 1860 to investigate heat and radiation.

Early thermometers used liquids expanding inside slender glass tubes and had scales which were simply equal length divisions ruled along the tube. It is easy to see that there was not necessarily a relationship to others made elsewhere.

In the late 17th century, it was realised by several people that the transition points where ice melted to water and water boiled to steam could be used as invariant standard points for a temperature scale. Newton is said to have used human armpit temperature as an invariant point on his scale and the familiar, but now deprecated, Fahrenheit unit (devised by Daniel Fahrenheit in 1710) had 100 degrees chosen to range from the coldest laboratory temperature known at the time to body temperature. In 1742 Celsius, of Uppsala, proposed the scale which designated 0 deg. as the icepoint of water and 100 deg. as the boiling point. (The Centigrade scale was independently suggested a year later but is now correctly called only Celsius). The remaining problem was how to subdivide accurately between the two fixed points — there were as many values at the end of the 18th century as there were interpolation thermometers. Liquid-in-glass thermometers are still most useful in everyday measurements, and as sub-standards, for nothing has been devised which is as reliable and inexpensive. For standards use, however, they have been overshadowed by superior methods.

Boyle's law states that the product of the volume and pressure of a given mass of gas is constant at constant temperature. The law holds well for the gases that are hard to liquefy such as helium, oxygen and nitrogen. Boyle's contribution led to the Charles'-Gay-Lussac law (they each had a hand in its discovery in 1787). This shows that the thermal coefficient of expansion of all gases is the same, provided they are held at constant pressure. This provided a more

tp — triple point bp — boiling point fp — freezing point	1927 ITS-27 °C	1948 ITS-48 °C	1948 IPTS-48 °C	1968 IPTS-68 °C	 K
tp hydrogen				−259.34	13.61
bp hydrogen, 25/76 atm.				−256.108	17.042
bp hydrogen				−252.87	20.28
bp neon				−246.048	27.102
tp oxygen				−218.789	54.361
bp oxygen	−182.97	−182.970	−182.97	−182.962	50.188
fp water	0.000	0			
tp water			+0.01	+0.01	273.16
bp water	100,00	100	100	100	373.15
fp zinc				419.58	692.73
bp sulphur	444.60	444.600	444.6		
fp silver	960.5	960.8	960.8	961.93	1235.08
fp gold	1063	1063.0	1063	1064.43	1337.58
fp tin		231.9	231.91	231.9681	505.1181
fp lead		327.3	327.3	327.502	606.652
fp zinc		419.5	419.505		
bp sulphur				444.674	717.824
fp antimony		630.5	630.5	630.74	903.89
fp aluminium		660.1	660.1	660.37	933.52

Fig. 2. The temperature scale is defined by several fixed points. This summarized table shows the minor changes made as measurement improved.

promising standard thermometer principle, for the actual gas used was of little consequence. Furthermore, the expansion coefficient of gases exceeds that of liquids and solids. In the early 19th century, air-filled thermometers were the accepted standard, as scientists could construct their own and obtain the same results.

In an effort towards further improvement, scientists of the time again turned to thermodynamics and it was Lord Kelvin who realised that Carnot's heat-engine cycle held the key to an absolute thermometric scale. The Carnot cycle is the most perfect thermal cycle that can exist in a heat engine. The Otto cycle is a less perfect one used in some internal-combustion engines and the Stirling cycle is another. Each of these are practical realisations of ways to convert fuel energy into mechanical energy via a thermal process. In essence, Carnot's theory states that the maximum efficiency of energy conversion occurs when the temperature of the energy input medium (be it steam, petrol combustion, etc.) at each stroke is made to vary from the hottest available at the commencement of the cycle extracting the heat, to the coldest possible, that is, absolute zero. This is logical, for no heat can be transferred from an object at absolute zero, as nothing can be made colder. Absolute temperature can, therefore, be defined theoretically and Kelvin defined his absolute, or thermodynamic, scale with zero at the absolute zero. He also arrived at the value of 273 degree K (K is the symbol for absolute temperature) for the ice point. The thermodynamic scale is easily related to Celsius values, for

Fig. 3. Establishing the oxygen point (−218°C) on the IPTS-68 scale at the National Measurement Laboratory in Sydney.

only the value 273 has to be added to Celsius values. He chose fixed points along the scale to coincide with various transitions of state. Kelvin's scale is identical with the gas thermometer scale.

In 1927 it was internationally agreed that six reproducible equilibrium states be used to define the absolute scale and Celsius values were assigned to each. This is known as the

International Practical Temperature Scale, IPTS for short. In 1948, and again in 1968, small changes were made in the values of the definitions as can be seen in the table of Figure 2 where the main features are shown. In general, the normal user of thermometers does not become involved in lengthy standardisations – (one equipment is shown in Figure 3). Instead we have secondary standards (calibrated thermometers of various kinds) that are used only to calibrate everyday instruments. Interpolation between the IPTS fixed points is made with defined electronic thermometers which are able to provide the continuous scale division required. Several methods are used – the resistance thermometer covers the range up to 630 deg.C, thermocouples from there to 1064 deg.C and above that the indirect radiation method of pyrometry takes over.

TRANSFER OF HEAT

A proper understanding of temperature measurements (do not forget measurement must precede control, and control is only as good as sensor accuracy) requires a knowledge of how heat is transferred. This gives clues to obtaining accurate measurements and provides us with principles that can be invoked to transduce temperature. It is by looking at the fundamental principles in new ways and in new combinations as technology changes, that our state of the measurement art is improved.

Heat transfers by conduction, convection or radiation processes. Conduction is explained by the kinetic energy principle, in which adjacent particles impart energy to those having lower energy. The energy exchange process continues until thermal equilibrium is established. The rate at which conduction takes place is decided by the thermal conductivity of the material and this parameter is most important in temperature measurement and control.

If a slab of material is heated at one end, the rate of heat flow is dictated by its thermal conductivity and the temperature difference between the ends. Thermal conductivity can be regarded as analogous to electrical conductivity, as can heat flow to current, and temperature difference to voltage. Consequently, the solution of thermal conductivity problems can be treated in much the same way as simple series electrical circuits. A fact often overlooked, however, is that the interfaces between layers may have considerably greater thermal resistance than the layers themselves. By way of an example, when air is flowing in a steel pipe at room temperature and atmospheric pressure, the tube has 100 arbitary units of thermal conductivity, the air-steel interface 20 and the air to internal probe 0.02 units. Consequently, when a thermometer is used it must be ensured that the active part of the device is truly at a close enough temperature to the body or fluid being measured. Silicon grease is a good heat conductor and is used to bed transistors onto heat sinks and thus obtain good thermal contact: a similar practice is recommended when bedding temperature sensors. Sometimes the sensor dissipates heat. In this case the indicated temperature will be high, as the sensor will be at a higher temperature than the body being measured, due to encapsulation conductivity. Another source of error could be that the thermometer mounting is conducting external heat to, or away from, the sensor. Simple basic rules for such cases are shown in Figure 4a.

Convection occurs in substances that can flow: heating reduces density and colder sections of fluid therefore settle displacing the hotter. If there is no physical medium, convection cannot exist. If, however, the temperature gradient in a medium is sufficiently low, convection will cease producing

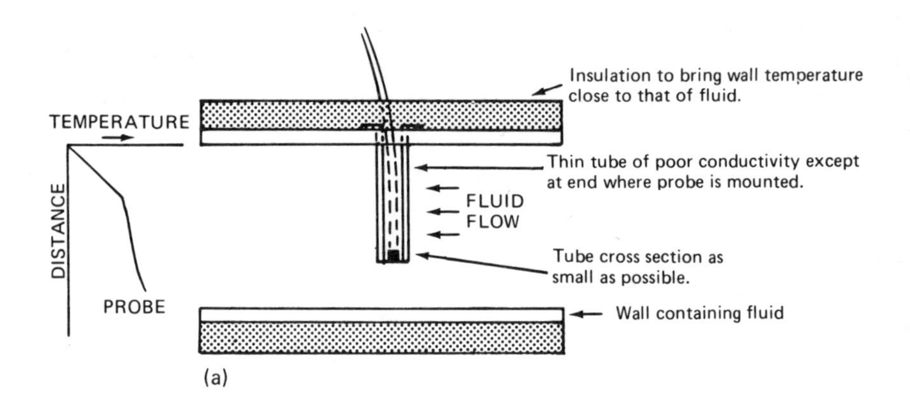

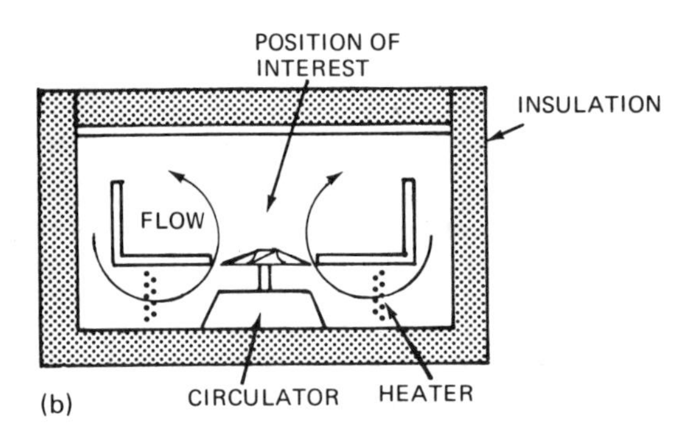

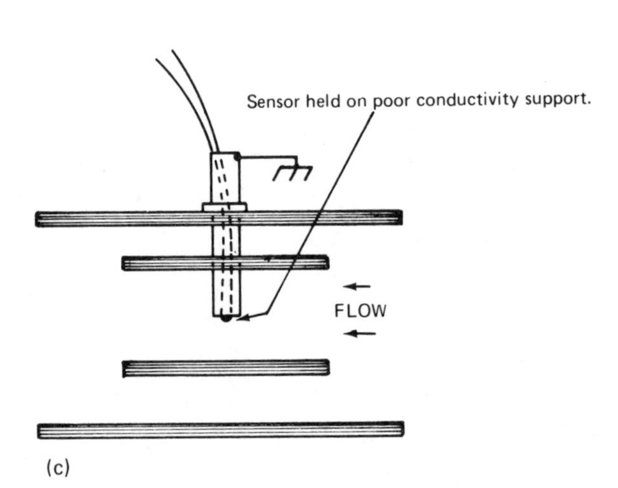

Fig. 4. Mounts for thermometer sensors need good thermal design to reduce errors due to thermal transfer mechanisms.

a) points to be observed when installing a thermometer well.

b) a good temperature-controlled bath circulates the fluid to reduce convection and conduction errors.

c) shields can be used to reduce radiation errors.

Fig. 5. Two metals having different expansion coefficients are often used to provide a dimensional movement from a temperature change.

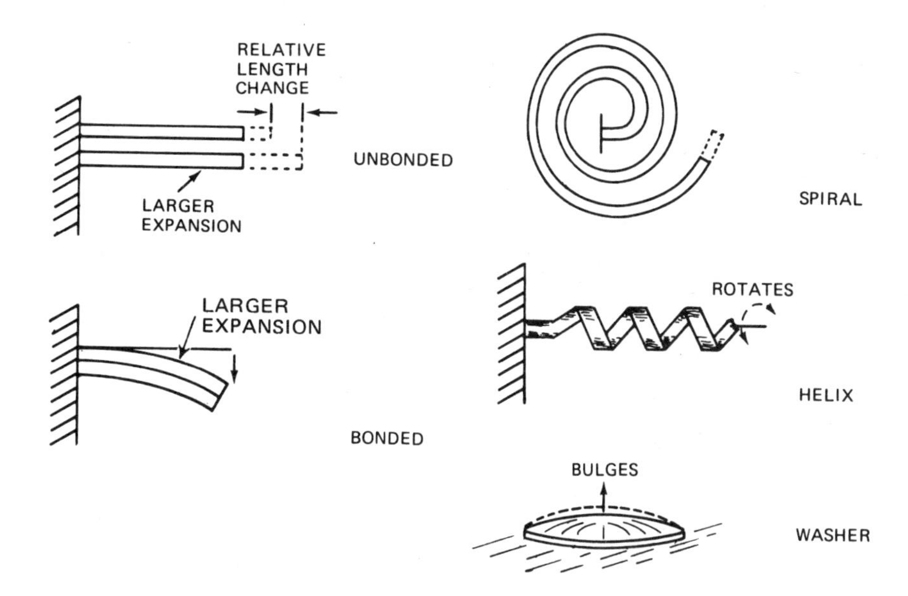

what is called inversion. Again sensor location must be chosen accordingly. In a car engine the temperature gauge sensor is placed high in the block to ensure a conservative reading. Fluids with poor thermal conductivity should be stirred, Figure 4b, to obtain reliable temperature measurements.

The third form of transfer, radiation, transmits heat by virtue of electro-magnetic (EM) radiation. More will be said of this in the following Chapter where non-contacting methods are discussed. Radiation is most effective in vacuum for the EM waves are not attenuated by any interposed medium. The degree of transfer depends much upon the surface characteristics and is greatest when irregularities are of the same size as the radiation wavelengths. Shields are sometimes used around a sensor, as in Figure 4c, to ensure that the thermometer is measuring the medium and is not influenced by external radiation.

The measurement of temperature is as old as dimensional measurements; there are, therefore, many methods in use today. They can be grouped into expansive, direct electrical, radiation and acoustic techniques, in the main, with a large group of miscellany which sometimes provide answers for unusual problems.

EXPANSIVE DEVICES FOR MEASURING TEMPERATURE

Of the many means by which temperature changes affect physical things, the most obvious is movement due to expansion and contraction. It is, therefore, not so surprising that the earliest thermometers operated on this principle with water, alcohol and, later, mercury in glass tubes.

In expansion thermometers, operation depends on the relative coefficients of two materials, which may be solids, liquids or gases in various combinations. The indicated temperature is not absolute, so standardisation is needed. We shall see that some methods *are* absolute producing an output that is directly related to temperature.

BIMETALLIC SENSORS

If an attempt is made to measure the change in length of a rod due to temperature effects, it will be found that the standard which is used for comparison may be altering in length also. Ideally, the standard should be made of a zero thermal coefficient material or be temperature controlled during the measurement. To avoid these requirements, the relative length change between two bars of different coefficient material is monitored rather than the complete length of each. In the bonded bimetallic thermostat, two strips of different metals are bonded together. If each has a different thermal coefficient, the strip must bend as the temperature changes. Ideally, the two should have the greatest possible difference in expansion rate, implying the use of the largest positive and the largest negative coefficient materials. Few satisfactory negative materials exist, so conventional designs use a combination of the smallest and largest positive materials — invar and steel or brass alloys usually. A variety of shapes are used as illustrated in Figure 5 — flat strips, coiled helices, disk washers and flat spirals. In each case temperature change produces a linear or rotary movement which is used to actuate contacts or drive a microdisplacement transducer (if a continuous output signal is needed). Although these methods are generally regarded as useful to a precision of only 0.1 deg.C (due to marketed versions being made to perform only to such limits), they are capable of extreme sensitivity if coupled to secondary transducers. Their simplicity, low cost and range capability, from −50 to 500 deg.C, makes them an obvious first choice for temperature control. The same principle is often used to produce a mechanical movement to compensate for temperature errors in an instrument.

LIQUID-IN-GLASS THERMOMETERS

Early thermometers used water in a fine glass tube was open at one end. It was soon realised that the varying air pressure seriously influenced readings and now the top space is filled with an inert gas at sufficient pressure to keep the liquid thread continuous. It is the relative volume change between the glass (steel sometimes) container and the liquid that causes the liquid to move in the graduated tube. Thermometers must, therefore, be used in the same conditions of immersion (bulb only, thread and bulb or complete unit) in which they were standardized.

Mercury is not the most sensitive liquid, toluene expands in volume at a rate five times that of mercury, and methyl alcohol at seven times. It is, however, particularly useful in contact thermometers where the mercury thread completes a circuit with an internal contact and hence, for example, cuts off the heater in a control system. The difficulty with liquid expansion thermometers is that range must be reduced with increasing sensitivity. Special designs are used, such as the Beckmann thermometer, in which the set point can be varied by manipulating the quantity of mercury in the reservoir, placing that unwanted in an upper storage reservoir. These can be read to 0.001 deg. C within a

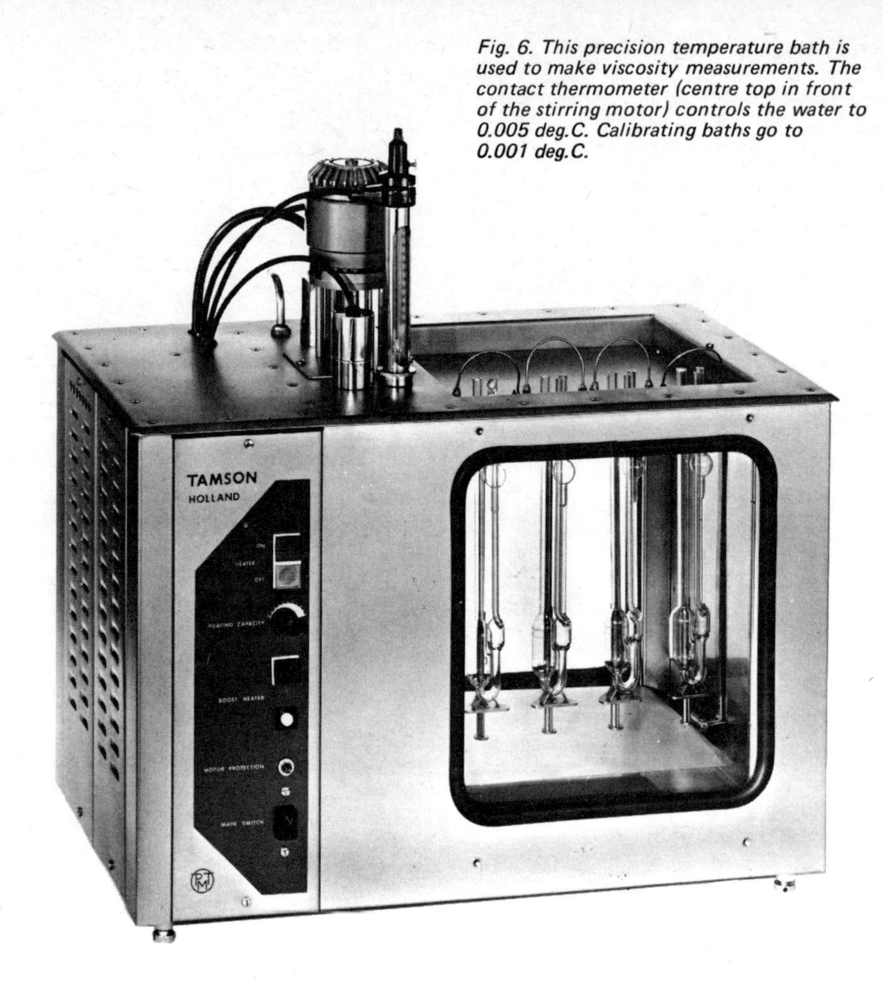

Fig. 6. This precision temperature bath is used to make viscosity measurements. The contact thermometer (centre top in front of the stirring motor) controls the water to 0.005 deg.C. Calibrating baths go to 0.001 deg.C.

range of 5 deg. and contact thermometers sensing to 1 millideg.C have been used. A precision, controlled temperature bath is shown in Figure 6.

A disadvantage of the liquid-in-glass thermometer is its zero depression characteristic. When subjected to a temperature change, the glass volume changes as does the liquid. The glass, however, takes a finite time to regain its original volume so a movement back toward zero from a higher temperature causes a depression of the zero which can last several hours or even days, depending upon the sensitivity of the thermometer. The effect is not large — modern glasses give rise to 0.01 deg.C error per 100 deg.C change.

Although mercury-filled manometers (devices for measuring pressure by the height of a column of mercury) have been automated to provide a readout of thread position as an electrical signal, it is surprising to find that no commercially available, proportional-output mercury-thermometer is marketed.

Also operating on the expansive principle are pressure thermometers in which mercury or xylene completely fill the system under an initial pressure. Heat at the bulb causes the internal pressure to rise operating a pressure sensitive element such as a Bourdon tube or diaphragm which is electrically instrumented. Some car temperature gauges and most radiator thermostat units operate this way, the latter illustrating a neat solution where electronics could not provide as simple an answer.

QUARTZ CRYSTAL THERMOMETERS

The resonant frequency of a quartz crystal depends upon temperature. Temperature changes crystal dimensions and the velocity of acoustic waves within it.

In operation (the Hewlett Packard version is shown in Figure 7) the frequency is measured by a timer counter and displayed directly in degrees by a digital readout. The advantages of the method are its high degree of linearity (±0.05%) (as good as the best platinum resistance thermometer) and resolution. In the short term, it is possible to discern 100 μdeg.C changes; over longer periods 0.01 deg.C.

Fig. 7. The Hewlett-Packard quartz crystal thermometer with digital readout can monitor temperature to 100μ deg.C.

DIRECT ELECTRICAL METHODS

All methods discussed so far make use of heating effects to provide a mechanical displacement, to which a secondary electrical output device could be attached. Several direct, that is, temperature to electrical signal, methods exist. These can be grouped as resistance, thermoelectric, thermistor and semiconductor junction thermometers. Thermistors are a form of resistor but their differences are distinct enough to place them in a category of their own. The first two methods date back to the early 19th century, the others to just a decade or two ago, for they are products of the semiconductor age.

RESISTANCE THERMOMETRY

Although Sir Humphrey Davy had realised in 1821 that the resistance of metals generally increased with increasing temperature, it was not until 1871 that this observation was put to use for temperature measurement. At that time Sir William Siemens constructed a special Wheatstone bridge for measuring changes in temperature of the resistance sensing element in his radiation detecting instrument. The modern form of resistance thermometer is due to Callendar and Griffiths who published a work in 1887 upon which today's procedures have been largely based. Resistance thermometry is still the most reliable method available, (along with thermocouples) being simple to use, having a good degree of linearity over a wide range and, not the least, assured long-term stability. It is not as sensitive as modern methods however.

In practice, the temperature sensing resistance is placed in the environment to be measured and its value read using a modified Wheatstone bridge. The resistance values are necessarily low (100Ω is typical) so connecting lead resistances are comparable in value. For this reason in accurate work, special bridges, see Figure 8, are used in which extra extension leads are connected to the sensor element to balance out the effect of leads. The three lead bridge (Callendar-Griffiths) suffices for general industrial plant control (potentiometric resistance recorders have three lead connections built in) but for standards work, a Mueller bridge (devised in 1939) is used. To give an idea of current sophistication, bridges can now be used to measure the hundred ohm value to 1 μΩ. Special precautions, such as the use of ac excitation with reactance instead of resistance elements, and temperature control of components and switch contacts to 0.01 deg.C are employed.

Platinum is the metal used for sensors in IPTS work. This was chosen because of its immunity to chemical corrosion, stability of resistance, high melting point and reasonably high specific electrical resistivity (ten times that of copper but far less than many resistance wires). Extreme care is exercised when winding the typically 100 μm diameter wire onto stable formers as mechanical stress in the element could produce a strain gauge rather than a temperature sensor. The spiral is then housed in a tube filled with dry air to prevent contamination. Sensors are available in all shapes and forms and the cost is reasonable.

The temperature scale indicated by the resistance change of a platinum thermometer does not agree entirely with the IPTS fixed point scale. To overcome this, correction values are calculated using equations that have been developed to convert measured values into the desired IPTS values. This is the world standard for laboratory temperature measurement from −270deg.C to 660 deg.C and has also found extensive day to day use in less sophisticated forms where ultra-precision is not required. To speed up the calibration of service instruments, a sub-standard transfer thermometer is intercompared with one requiring calibration using a controlled temperature bath (like that in Figure 6) which is filled with water, oil or salt as the temperature demands.

Platinum sensors are reliable to at least 0.01 deg.C as far as IPTS reproducibility goes, but in cases of control, where the absolute value is not critical, they can do far better. Probably the best example to date was in a calorimetry bath built by the National Bureau of Standards (NBS) in 1968. A most sophisticated control system using a commercial platinum sensing element held water at the centre of the container to within 20 μdeg.C per day.

Not all resistance thermometers use a wire wound element, for that form of construction is often too large and has insufficient thermal response. For example, in the United States, several people have built sensors made from 1 mm of 630 nm diameter platinum wire to measure millisecond temperature variations in the atmosphere. Here the mounting is critical, to minimize external heat conduction to and from the sensor. Vacuum deposited films have also been employed.

Increasing interest in cryogenic temperatures (those near absolute zero) has resulted in methods specially suited to that region. One simple solution uses the common carbon composition resistor as the sensor. Calibration equations are needed to correct the actual readings in the 0.6 −5 K region to 0.001 K precision.

THERMOCOUPLES

In 1821 Seebeck discovered that when dissimilar metals were joined together to form a circuit of at least

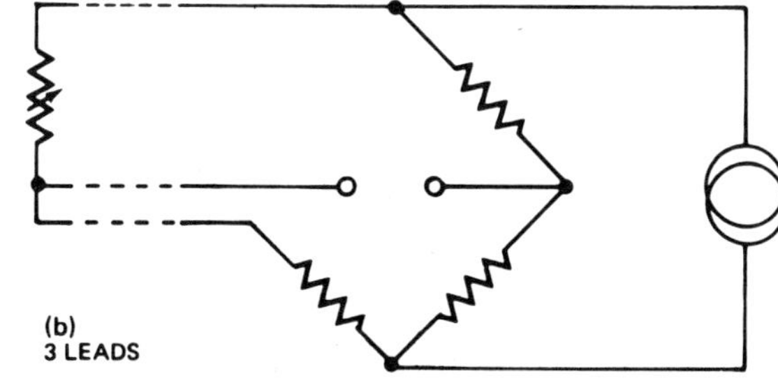

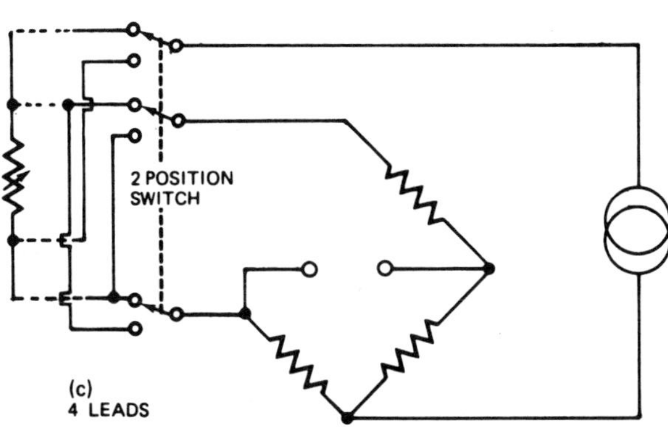

***Fig. 8. Resistance** sensors are used with extra extension leads to cancel out the errors caused by lead resistance. a) In a conventional Wheatstone bridge lead resistances add to the sensor. b) The Callendar-Griffiths bridge uses two leads from one end of the sensor thus balancing out lead resistance. This is commonly employed in industry. c) To eliminate all lead errors this Mueller bridge is used. The two readings are averaged.*

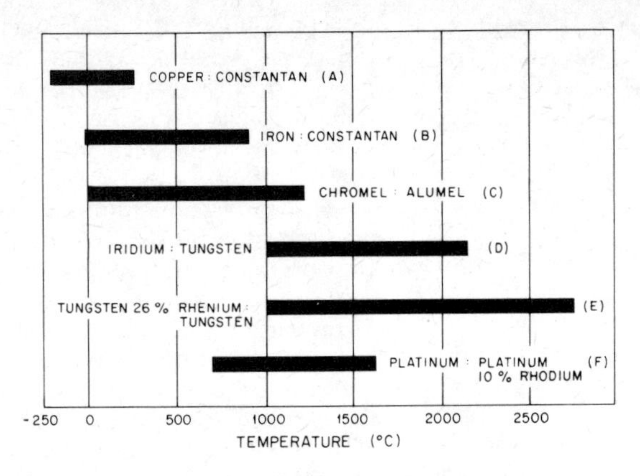

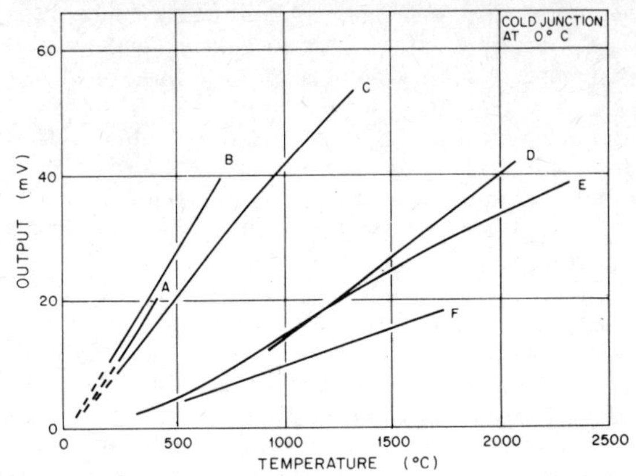

Fig. 9. Ranges and output-temperature relationships for commonly used thermocouples.

two junctions, a current would flow when the two junctions were at different temperatures. Peltier, in 1834, observed that the reverse also applied – passing a current through the loop caused one junction to cool, the other to heat. The Seebeck voltage is small being of the order of tens of microvolts per degree so the method is comparatively insensitive. By contrast thermistors can produce signals of millivolts per degree. A chart of the voltage output for common couples is given in Figure 9. Provided one of the junctions (the reference couple) is held at a constant temperature, temperatures measured with respect to it are absolute for no calibration is needed if the materials of the couples are known. Tables of values are available for the commonly used combinations which will enable the thermovoltage to be converted in IPTS values.

Thermocouples are formed by joining two wires together (twisted together often suffices, otherwise they are welded or clamped). Outputs vary from the couples and are not linear over the entire range (they closely follow a parabolic curve of which only the linear region is used). For this reason the sensitivity of a couple depends upon the temperature. Copper to constantan, for instance, gives 18 μV/deg C at – 183 deg.C and 62 μV/deg.C at 400 deg.C. A couple made from two similar materials of different batch, copper from two different coils of wire for example, can generate as much as 10% of the voltage of a recommended thermocouple so care is needed in wiring. Special leads are sold for connecting thermocouples as all connections are potential temperature sensors. For high temperature work, such as in furnaces, room temperature is adequate for controlling the temperature of the reference couple (temperature sensitive special resistors are usually included in temperature recorders to allow for ambient changes). For ambient temperature operation, however, the reference couple must be better controlled with either a simple ice bath (miniature controlled Peltier units are available) or a special temperature tracking power supply that simulates a couple at the ice-point.

The advantages of thermocouples are their low cost, extremely wide temperature range (with different materials they cover from absolute zero to several hundred degrees Celsius) but the main feature is often the small size possible which enables millisecond response times to be realised.

The simplest way to measure temperature by thermocouple is with a milli-voltmeter and a set of tables. Although the circuit draws current to drive the meter, the resultant Peltier effect is negligible. For more precise work, potentiometers are used to determine the unloaded voltage. It is possible to resolve to 10 nV with a good potentiometer so thermocouples are useful to 0.001 deg.C but great care is needed at such sensitivity to eliminate stray thermoelectric effects.

As with resistance sensors, the universal demand has caused manufacturers to provide a wide range of hardware for applying thermocouples. Self balancing recorders are available that display temperature directly provided the correct couple is used. In these the pointer/pen is driven by a powerful, but crude motor which also drives a coupled potentiometer. If the potentiometer output does not balance the thermocouple signal, the error between the two is amplified electronically and used to drive the motor accordingly. The system rapidly balances and by this expedient the recorder is made extremely robust whilst retaining full scale sensitivity of a few millivolts.

To increase the sensitivity, several couples (a thermopile) can be joined in series so that more than one couple is measuring temperature. There will, therefore, be more reference junctions also. For example, to test the NBS water bath mentioned earlier, a

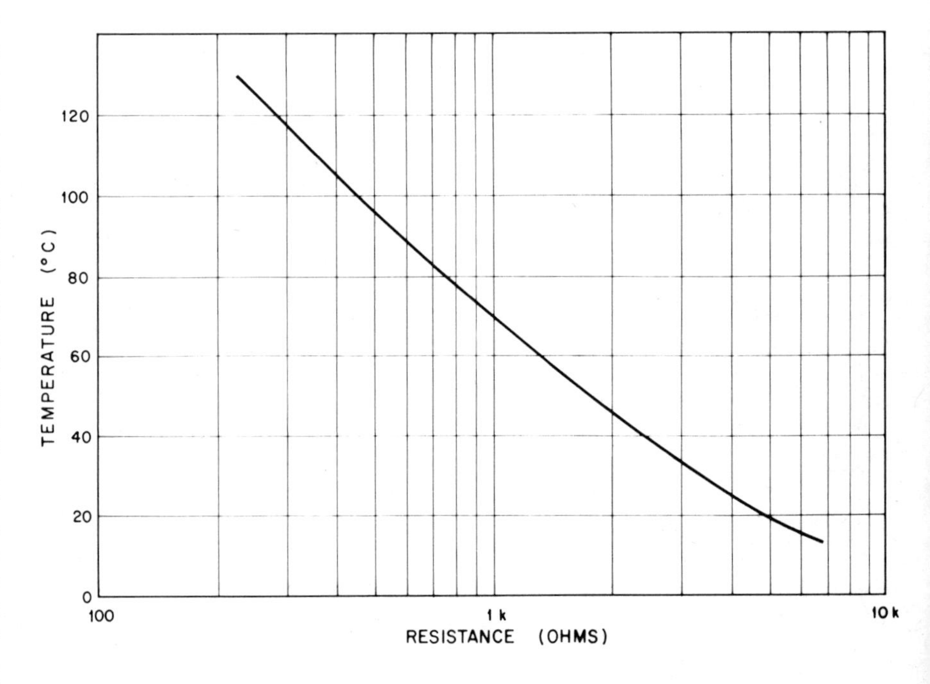

Fig. 10. Resistance variation with temperature of a typical thermistor.

thermopile was placed in the fluid and the reference pile in the centre of a fluid filled, large size, Dewar flask that provided a time constant of many hours to the reference temperature. By this means they were able to verify the short-term stability, the thermopile sensitivity being limited by electronic resistance noise to an equivalent temperature of less than 1 μdeg.C.

It was not until the mid 1940's that suitable temperature sensitive bulk resistance materials could be made with stable characteristics. In 1946 staff of Bell Telephone Laboratories reported their work on thermally sensitive resistors from which the name thermistor has derived. Thermistors usually are made from solid semiconducting metal oxides and have a large negative temperature coefficient of electrical resistance. (Positive coefficient thermistors also exist but are less common). Resistance is exponentially related to temperature, Figure 10, and two constants, which are quoted by the maker, enable the characteristic curve to be drawn to within a few percent. A wide range of nominal resistance values (usually quoted at 20 and 25 deg.C) are available — ohms to tens of kilohms. The temperature coefficient is greater than for conductor resistance sensors. For example, one commercial resistor that is 10 kohm at 0 deg.C is 153 ohms at 100 deg.C. but, of course, the relationship is highly non-linear.

Thermistors are used in much the same way as resistor sensors — that is, in a bridge network. Lead resistances are usually uncritical being orders of magnitude less in value than the thermistor value.

It is rare to see design procedures for thermistor thermometers in text books as yet but the process is reasonably simple (see the reading list). Factors to be considered are the choice of optimum bridge resistance values, for the sensitivity (in mV/deg.C usually) depends upon these as well as the thermistor type and applied voltage. Bridge currents heat the thermistor so it is necessary to design for a tolerable rise in offset temperature (the encapsulation insulates the sensor from the environment enough to produce a temperature drop across it); this fixes the maximum bridge excitation voltage. Finally, the sensitivity can be calculated. Typical values will lie between 1 and 100 mV/deg.C depending on the offset allowable. There is seldom real need to design for high sensitivity in the bridge as integrated-circuit operational amplifiers can provide adequately stable gain. For best results (but not always needed) the bridge can have ac excitation and use phase sensitive rectification to enhance the signal-to-noise ratio at the output.

Thermistors are available that can operate from cryogenic temperatures to well over 300 deg.C using three different groups. Encapsulations vary from dot size beads to flat disks. They are also available ready mounted, see Figure 11, in devices ranging from 500μm diameter hypodermic needles to the robust sensors used in car engines.

Portable electronic thermometers using thermistors (Figure 12) are now commonplace. When first introduced they were said to be unstable and unreliable with time but this is not so now. Well aged units (the manufacturer does this) can hold temperatures stable to 1 millideg.C per month and even better over shorter time periods. Their non-linearity can be reduced by using them in combination with series and parallel fixed resistors but at the expense of sensitivity. For wide range, accurate measurements, the resistance sensor still is superior but thermistors are finding more and more use each day. Electric motors use them (as well as

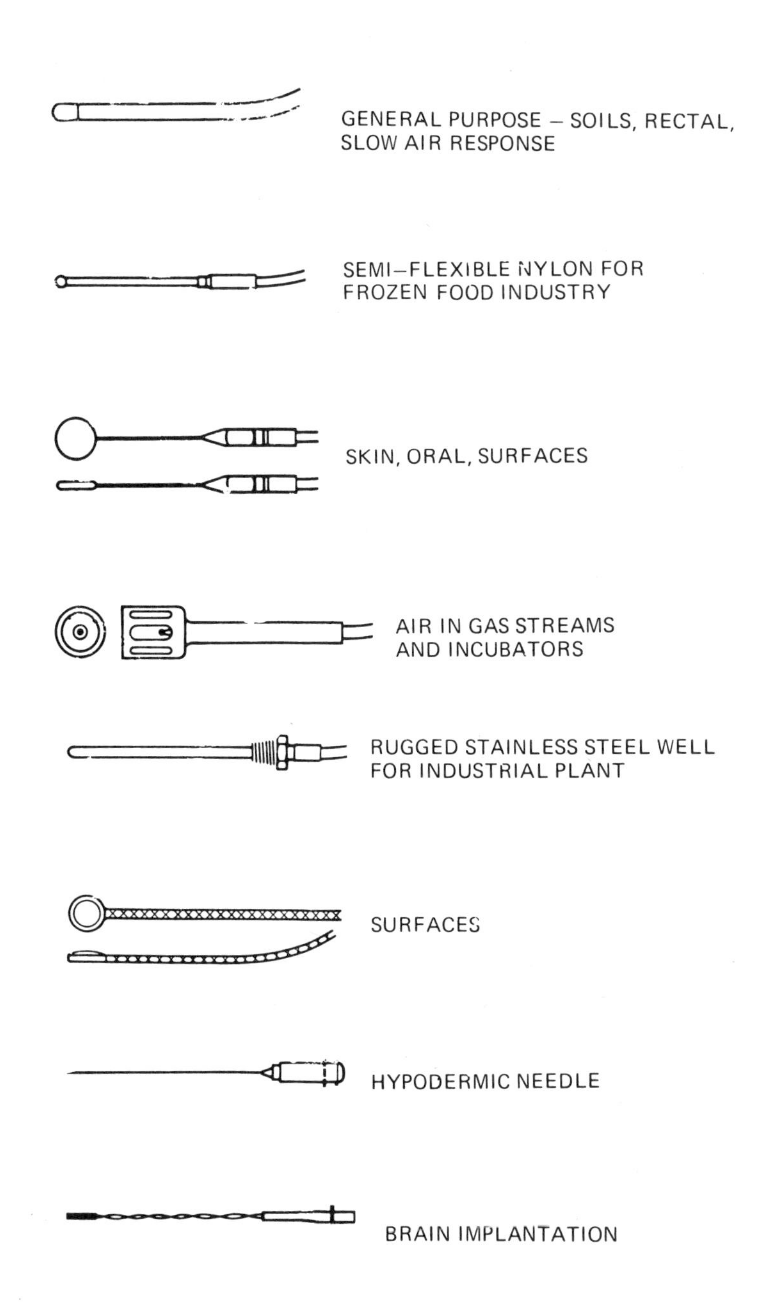

Fig. 11. Portable thermistor thermometers are now commonplace and inexpensive. The probes are available to cover most contingencies.

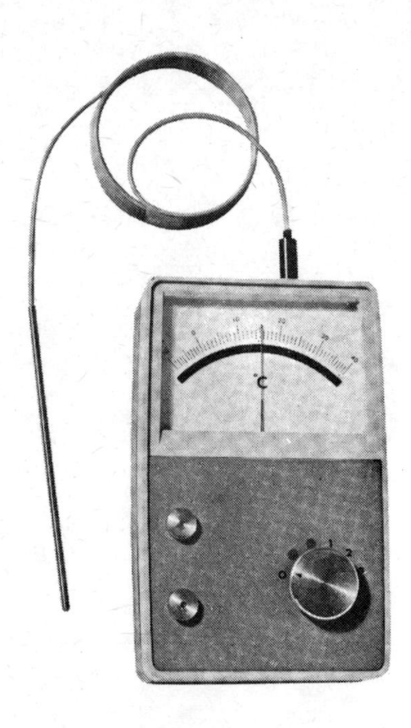

Fig. 12. Portable thermistor thermometer. (Grant Instruments).

resistance sensors) to sense the hottest temperature of the winding. This is obviously superior to bimetal devices that only monitor the average carcase temperature.

SEMICONDUCTOR THERMOMETERS

One of the basic shortcomings of early semiconductor devices, especially the germanium types, was their temperature dependence. The relationships between collector current, base-emitter voltage and temperature were known but it was not until 1958 that technical papers began to appear showing how to employ this defect for temperature sensing. Since then a few people have improved the technique to a point where a silicon transistor can be used to make an ultra-linear calculable thermometer for the range −50 deg.C to 100 deg.C in which the output signal is decided only by the temperature and knowledge of two basic physical constants. This method is not described in books on thermometry and therefore suggested reading is given at the end of this chapter.

If a silicon transistor is supplied with a constant collector current it can be shown theoretically that the base-emitter voltage V_{be} is proportional to the absolute temperature and is reasonably linear as shown by the curve of Figure 13. In this simple case Vbe is not entirely independent of the transistor materials or geometry. However, if the collector current is cyclically switched between two current levels it can further be shown that the output is now extremely linear (see Figure 12 again) and independent of all parameters of the device except for the ratio of Boltzman's constant to electron-charge (two precisely known fundamental numbers in physics). Sensitivity is reduced by the switching to 0.6 mV/deg.C compared with the approximately 2.0 mV/deg.C for the basic circuit having constant collector current. Linearity falls off above 150 deg.C and below −50 deg.C due to secondary effects becoming significant. Although complete circuit designs were published in 1962 and patents taken on features of the switched current method in 1968, there has been little interest in what appears to be a most useful thermometer principle. Several integrated-circuit manufacturers incorporate transistor junctions for controlling the chip temperature or shutting down the circuit in case of overheating but they do not use the V_{be} method. Like all methods it has a disadvantage. As V_{be} is proportional to absolute temperature the device output increases with temperature; at room temperature it delivers around 600 mV. If measurement is needed to high precision, say milli-deg.C, the expensive requirement of a precision digital voltmeter resolving to five or six decades is needed unless stable means of generating an offset voltage can be provided instead.

Thermocouples are similar to this — they generate the offset voltage with the reference couple that must be temperature controlled. By contrast, thermistor and resistor sensors, being passive devices, need only stable calibrated resistors (voltage supply variations are a secondary effect on precision) to make accurate measurements. So no matter which electrical method is employed, measurement precision is ultimately limited by the stability of a secondary physical component.

FURTHER READING

There are many books on thermometry; here is a selection. Few discuss thermistor or transistor methods at any depth so technical papers are also listed.

"Fundamentals of Temperature, Pressure and Flow Measurements". R.P. Benedict, Wiley, 1969.

"The Measurement of Temperature". J.A. Hall, Chapman & Hall, 1966.

"Methods of Measuring Temperature" E. Griffiths, Griffin, 1947. (This gives more detail of the older methods than modern texts).

"Measuring Temperature" L.C. Lynnworth and J.J. Benes, Machine Design. Nov. 13, 1969. 190-204. (An interesting summary of new techniques).

"Measurement Systems: Application and Design". E.O. Doebelin, McGraw-Hill, 1966.

"Precision Temperature Controlled Bath". E.C. Bell and L.N. Hulley. Proc. I.E.E., 1966, 133, 1667-77.

"p-n Junction (Transistor) as an Ultralinear Calculable Thermometer" T.C. Verster, Electronics Letters, 1968, 4, 9, 175-176.

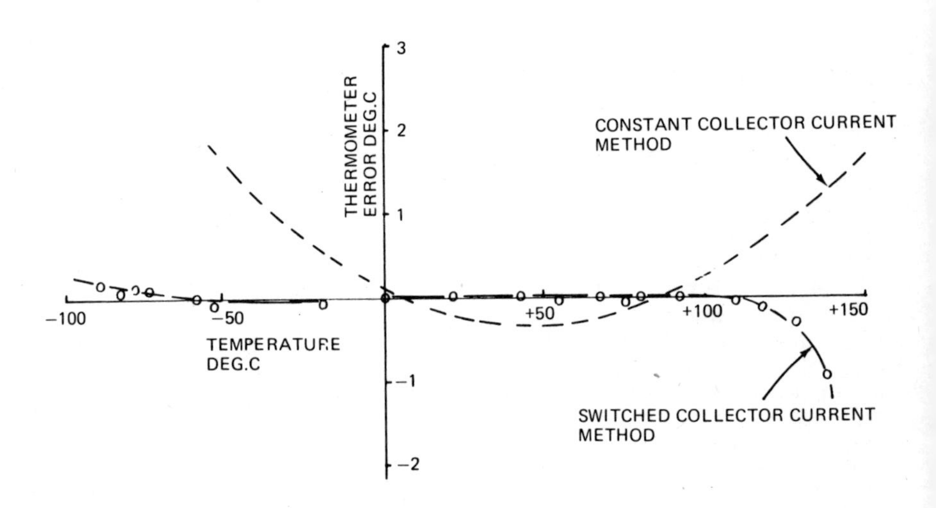

Fig. 13. Calibration curves for the transistor temperature sensor produced at the South African National Research Institute for Mathematical Sciences.

CHAPTER 7

NON-CONTACT AND LESSER KNOWN METHODS OF DETERMINING TEMPERATURE

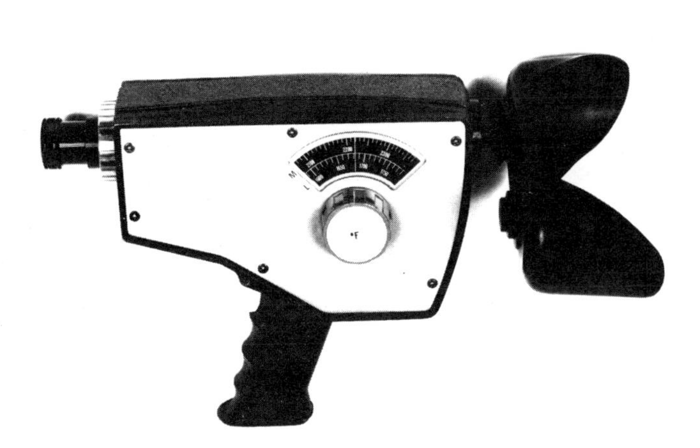

Self-contained lightweight optical pyrometer for measuring high temperature. (Leeds & Northrup)

The previous Chapter was concerned with scales, standards and the transfer of heat, all of which are common to every temperature sensing procedure. The introduction led to discussion of thermometers using the expansion principle and then to the direct electrical methods, namely, resistance, thermocouple, thermistor and semiconductor thermometry.

Another significant group of devices operate without need for mechanical contact with the medium to be measured — they make use of the *radiated* energy of substances.

THE RADIATION PHENOMENA

All physical objects radiate energy, the quantity depending upon their temperature and emissivity (the degree to which surfaces come close to ideal radiators). The presence of radiation must have been realised since antiquity, but only as visual experiences or feelings of warmth which were accepted without satisfactory explanation. Once it had been shown, in the late 18th century, that heat was in fact, energy, it was only a matter of time before quantitative explanation was evolved.

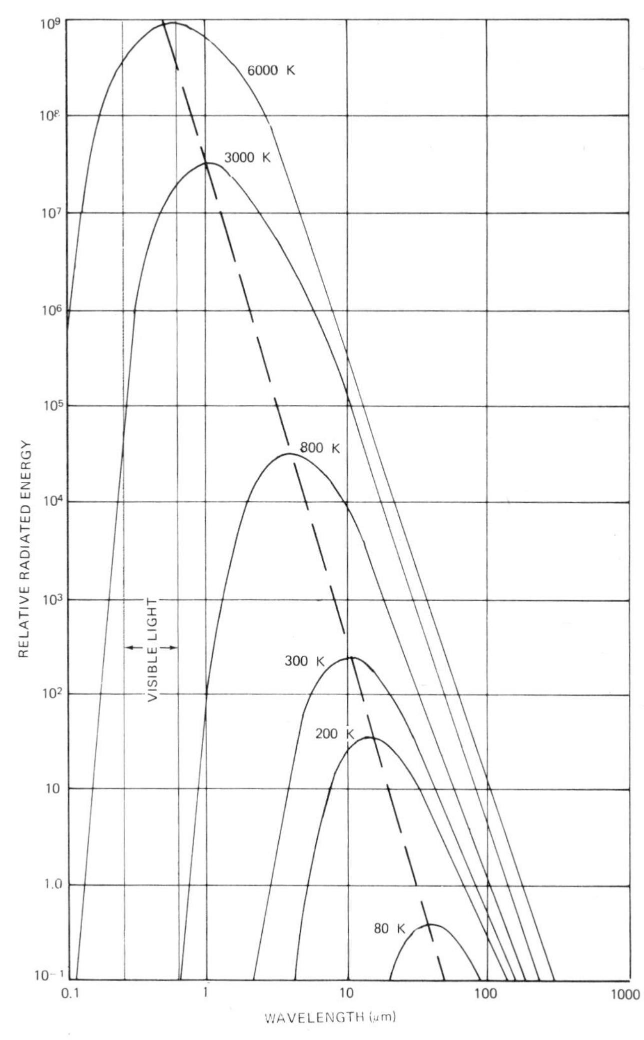

Fig. 1. Distribution of black body radiation as temperature varies. The total energy radiated at a particular temperature is represented as the area under the curve for each.

The relationship between the radiation from a surface and its temperature was first proposed by Stefan in 1879. His work, coupled with that of Boltzmann in 1884 led to the Stefan-Boltzmann law that states that the energy radiated from a unit area of surface in a given time is proportional to the fourth power of its absolute temperature. The law only applies for sources that are known as black body or full radiators. Such sources absorb all energy falling upon them and thus appear black to the eye. The law, however, only provides part of the picture for no information is available to explain where the broadband blackbody source-radiation peaks. A family of several curves, (Fig.1), shows the relationship between energy, temperature and wavelength for an ideal black body radiator.

It was a scientist called Wien who, in

1896, provided science with the law that describes how the peak shifts wavelength with temperature and how the energy is distributed across the spectrum at each value of temperature. A little later, in 1900, Max Planck suggested a modified expression that gave closer agreement with observed values over a wider range of wavelengths and temperature. However, for most purposes, Wein's law is adequate. As black body radiation enters into numerous scientific and industrial endeavours, tables have been computed for the many possible combinations and special slide rules made for calculating the total power, peak emission and distribution. For much of the need, however, simple graphs like Figure 1, suffice to estimate the energy within a given bandwidth. The actual energy received at the detector is found by subtracting the losses of the optical system (airpath attenuation, reflection and absorption losses in elements and detector efficiency) from the total known to be received at the entrance aperture of the instrument. If working over a long atmospheric path, the original radiation will no doubt be filtered by the transmission windows of the path. Finally, the detector output will be the transduced energy received through the instrument but modified by the spectral response of the detector. In other words, a detector should optimally be matched to have a significantly overlapping spectral response.

If an enclosure is uniformly heated, radiation issuing from a small hole will be very close to that from a perfect black-body. The actual substance from which the enclosure is made and the shape of the cavity matter little. An early source, used by Summer and Pringsheim in 1897 was a hollow sphere of copper, blackened on the inside. This was heated in a molten salt bath. It served for the study of radiation from 200 to 600°C. For higher temperatures they used a blackened iron cylinder heated by gas combustion. A modern source, used to establish the gold point of the IPTS scale (1063°C) has molten gold inside a cavity. Several manufacturers supply temperature-controlled cavities for standardising detectors in infrared work.

In practice some sources to be measured are close enough to the ideal black body for the laws to be used without modification. Looking into an industrial furnace is a good example, a hole bored to a depth of at least five times its diameter into the surface of an incandescent body is another near-perfect arrangement, for all energy entering such a cavity is internally reflected and absorbed at each reflection so that virtually none escapes.

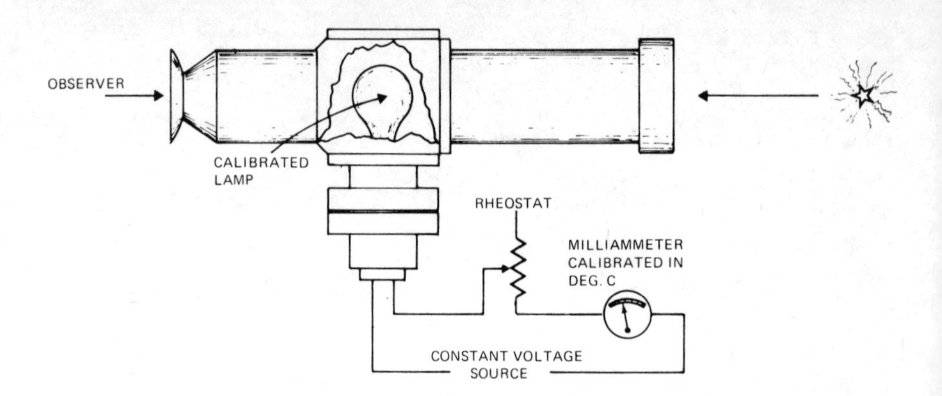

Fig. 2. The basic disappearing-filament optical pyrometer — the rheostat is adjusted until the light filament just disappears. Temperature is read from the dial.

There are, however, many instances (surfaces for example), where the appearance is not black and the emissivity is not unity but something less. A grey body is one similar in performance to a black body but where the emissivity is less than unity. Tables of emissivity of surfaces are available. By way of example, carbon can absorb as much as 94% of incident energy whereas unoxidised silver only 7%. This might lead one to think visual experience is a good measure of emissivity but that is not true for the eye has response over a narrow band of the black body radiation spectrum only. In fact, at infrared wavelengths the reverse situation applies. An object can never be brighter than the black body of the same temperature. Not all are black or even grey — some have an emissivity that follows no obvious law, perhaps peaking at various wavelengths, with zero at others. These are known as non-grey bodies. Gases do not radiate as black body radiation but rather at specific wavelengths. The temperature measurement of gases requires a different approach.

RADIATION PYROMETRY

There are two basic groups of devices by which temperature can be measured using radiation. These either determine the total radiation emitted or just the radiation in a narrow band of wavelengths. It is necessary in both cases to know something of their emissive behaviour, for all surfaces fall short of being perfect black body radiators. Temperatures calculated from the observed values will always be low to some extent; corrections will be necessary if precision is needed.

THE DISAPPEARING FILAMENT PYROMETER

Instruments used to determine the temperature of a source from its apparent brightness were formerly called pyrometers as they were developed mainly for furnace work. A more recently introduced term is radiometer but this term is also used for instruments used for measuring other forms of radiation than visible and infra-red.

Early pyrometers made use of the human eye as the detector of radiation. This biological photocell system is capable of only low grade accuracy when determining absolute radiation level but has quite high acuity when comparing two sources in the same field of view. In the disappearing pyrometer, (Fig.2), the eye sees the source (often through a calibrated filter to reduce the brightness to an acceptable level) and a heated filament in a common focal position. The filament current is adjusted until it just disappears into the background. In simple instruments, the current rheostat is calibrated directly in degrees, but in more precise cases the current is carefully determined with a standard resistor and a voltage measuring potentiometer. The value is then referred back to standardization curves obtained when the instrument was calibrated against sub-standard black body sources. Special tungsten lamps, usually with a strip filament, have been evolved for this purpose in order to retain their luminous efficiency over as long a period of calibration as possible. Such pyrometers can be intercompared by simultaneously viewing a common source — which need not be a black body.

Many processes need continuous operation and for these the balancing-out procedure has been automated. The cost of the instrument is naturally higher than for a manual device but often the total system application demands greater precision, faster response, improved reliability or continuous operation for measurement records and control, and all of these require the manual element to be eliminated. One technique alternately scans from a standard source to the surface of interest using a rotating or nodding mirror. A photo-cell determines if the brightnesses are equal and if not, the error signal is used to alter the source heater power accordingly. In recent years, a number of these instruments have been reported in the technical literature. A schematic of an early I.C.I. unit Figure 3 (used for monitoring fast moving synthetic fibre threads as they are made) illustrates the concept. A small heated aluminium plate provides a reference background temperature that, in this case, can be varied from 45° to 280°C. The nylon line moves between this source and the detector.

After collimation by a calcium-fluoride lens assembly (special materials are need for optimal operation at the infra-red wavelengths generated – there is little to be seen by eye at these temperatures) the image falls onto an infra-red photo-conductive cell. At the rear of the detector head is an eccentric cam that causes the unit to oscillate from the background to the background plus nylon-thread. The signals obtained are low level and noisy so synchronous detection is incorporated. The chopper wheel produces a higher fixed frequency signal from the cell. This is synchronously rectified in a unit called a phase-sensitive detector which produces the error signal which is used to control the heat of the source until balance is achieved. The temperature is measured as the power level to the background source.

One difficulty in pyrometry is that the emissivity may be an uncertain quantity and results subsequently imprecise. An approach to overcoming this, uses two measurements made at two different wavelengths within the broadband radiation. The principle, (hopefully invoked), is that the energy radiated at one wavelength increases with temperature at a rate different to that at another. Opinion differs on the effectiveness of this method. Benedict (in his book 'Fundamentals of Temperature, Pressure and Flow Measurements – see bibliography in the last Chapter) suggests it is rarely helpful, but research workers of the Central Electricity Generating Board in Britain have made use of it in a system for monitoring the surface temperature of a captive pulverized coal particle (of 0.25 mm diameter). Their application was to observe the events leading up to combustion as the coal is heated using the heat from focused lamps. This difference in outlook illustrates how instrumental methods can be condemned by a personal experience which may not have been adequate. The design of instruments is so incredibly fraught with unknowns and compromises that reasons for the lack of success can often only be found by extensive and perhaps prohibitive extra research.

Until quite recently, pyrometry was useful only for measuring high temperatures for, as can be seen in Figure 1, the total power radiated falls off as the fourth power as the temperature decreases. Traditionally, the pyrometer served as a means to measure visibly hot objects. A white-hot tungsten lamp peaks at 3000 K but, by reference to Figure 1 again, it can be seen that very little of the energy is actually radiated at visible wavelengths. Visual experience of radiation virtually disappears at around 800 K but there is still energy in the infra-red regions and this can be sensed as warmth. At absolute zero, no radiation occurs but as virtually nothing can be maintained at zero temperature, all objects emit radiation. The lower the temperature the longer the wavelength and the smaller the power level.

INFRA-RED RADIOMETERS

The infra-red portion of black body radiation has been the subject of scientific interest since it was discovered but little practical use was made of it until the 1940's when military scientists developed heat-seeking devices during the Second World War. Since then IR technology has improved enormously, especially in the area of detectors.

Visibly responsive photo-detectors such as the common silicon photocell

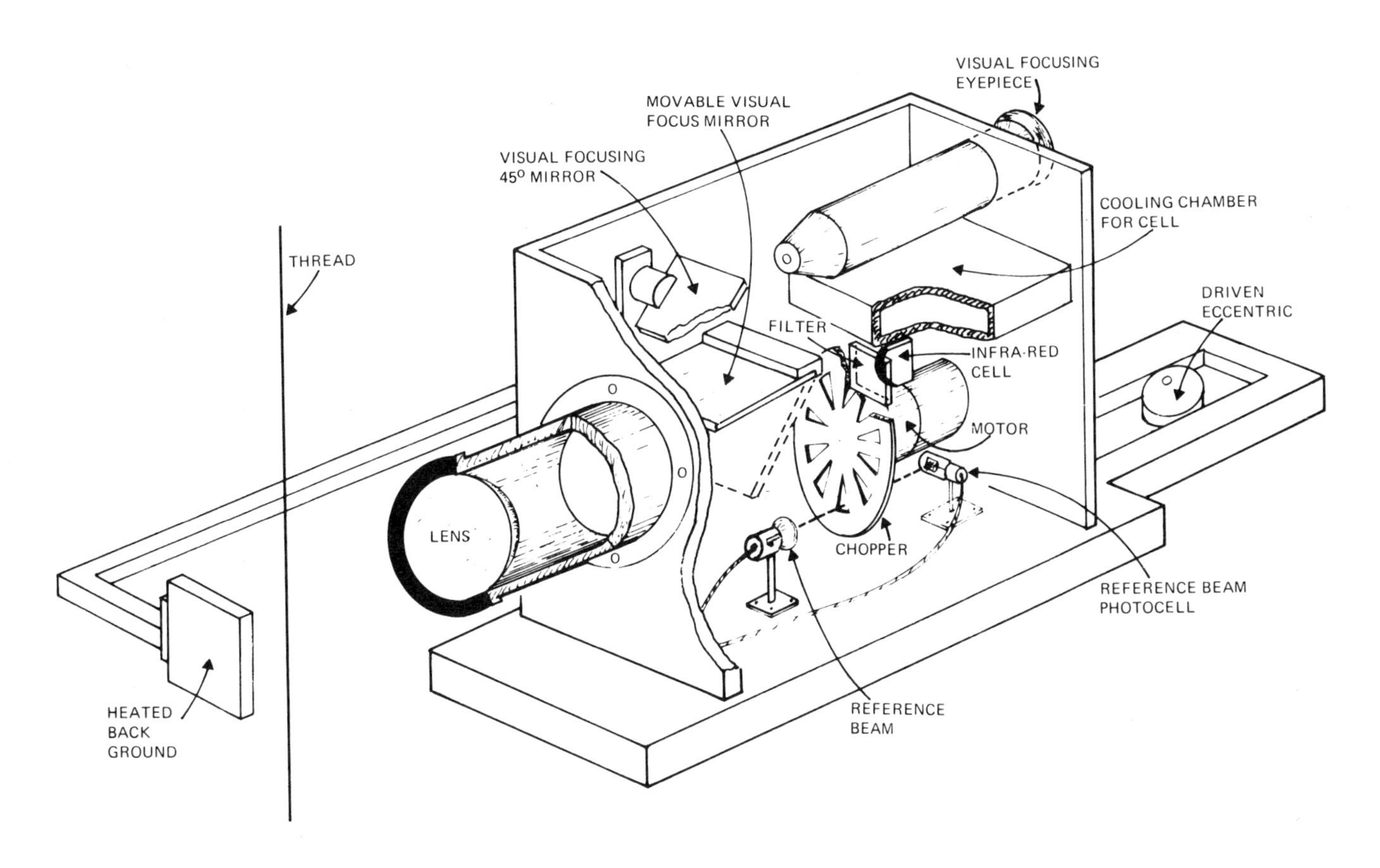

Fig. 3. Schematic of the I.C.I. nylon thread non-contact thermometer.

have some response in the near-infrared region but little as the wavelength increases. Early IR cells used the impinging heat to alter the resistance of a fine wire or, later, deposited films — these are known as bolometers. They are characterized by broadband response but lower sensitivity.

A chart of detectors is given in Figure 4; and as can be seen, apart from bolometers, no device covers the spectrum. Because of this detectors must be chosen to suit the wavelength of interest.

Detectors with peak response at 35 μm wavelength are available, so it can be seen that temperatures down to a few degrees absolute can be detected provided the noise level is controlled. Much of the noise seen by a detector is generated as emission from its own components and adjacent mounts. Furthermore, at room temperature the electrical resistance noise of the element can swamp the signal. For sensitive IR radiation detection, the detector is usually cooled by operating it on the bottom of a specially designed Dewar flask filled with liquid nitrogen or helium or, for continuous operation, on the end of a miniature cryostat. One important parameter of the design of IR radiometers is the cost of the detector — for they can be very expensive.

The sensitivity of IR detectors is now universally quoted as the specific detectivity (D*). This is a normalized value of sensitivity that makes allowances for different active element areas, electrical bandwidth of operation and the measured noise equivalent power (NEP is the rms infra-red signal incident upon the detector that produces a unity signal-to-noise ratio). All detectors can be intercompared on this common basis. Without becoming involved with actual D* figures, it suffices to say one reported modern IR system can sense the detail of a television tower through 70 km of thick fog.

The basic IR radiometer is very similar in principle to the thread temperature measuring pyrometer already described, the main difference being that the temperature reference is not placed behind the object, for that is usually not possible. Instead, a small black body source is built-in so that it and the scene are viewed alternately.

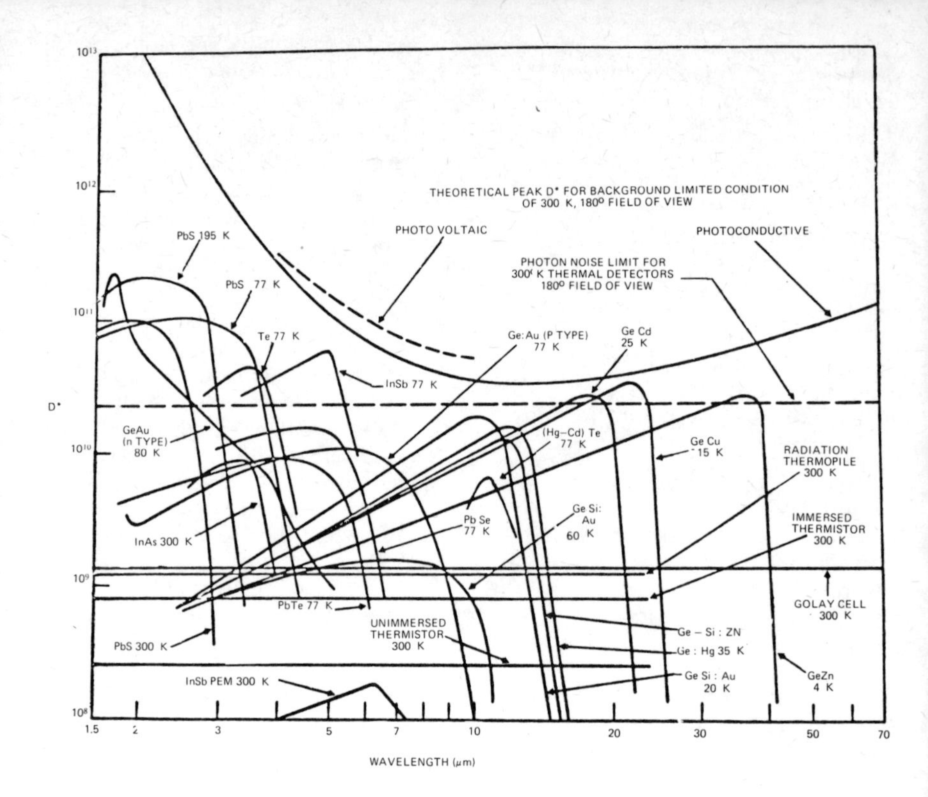

Fig. 4. Relative response of photodetectors operating over the visible to far infrared spectrum.

The relatively low power levels involved with low temperatures means that the IR radiometer must have a high gain telescope to gather as much energy as is practicable. In the unit shown in Fig. 5, a folded reflector telescope is used. Refracting elements are to be avoided in IR work (unless small) for the cost of the exotic materials needed rises rapidly with size. Reflectors, on the other hand, require only low-cost reflective coatings. The entrance aperture to the telescope, however, needs a transmitting cover but some simple expediencies, such as using thin plastic sheet helps — in this case the focusing properties are of no importance.

A chopper system (perhaps a rotating wheel or vibrating slit) is incorporated to enable synchronous detection to be used. This can allow signals as small as 100 dB below the noise level to be recovered. In the system illustrated, the chopper also serves to switch the detector to the scene and reference source alternately. Many rocket probes have carried radiometer devices like this into space to determine the emission and transmission losses of the sky and atmosphere. Narrow bandpass optical filters (only nanometre bandwidths at times) are usually employed to select the wavelengths of interest along with multiple detectors with different peak responses.

THERMAL IMAGING

Initial interest in IR was for the detection of the hot exhaust plumes and engine covers of aircraft, the

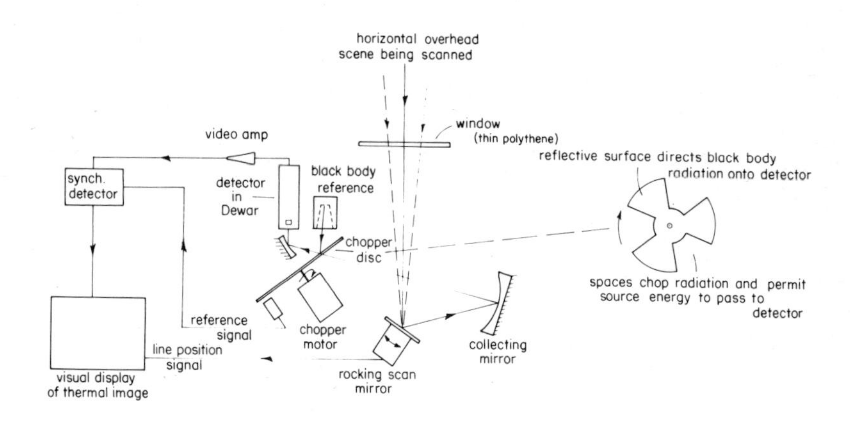

Fig. 5. Basic block diagram of a radiometer used in the infrared regions.

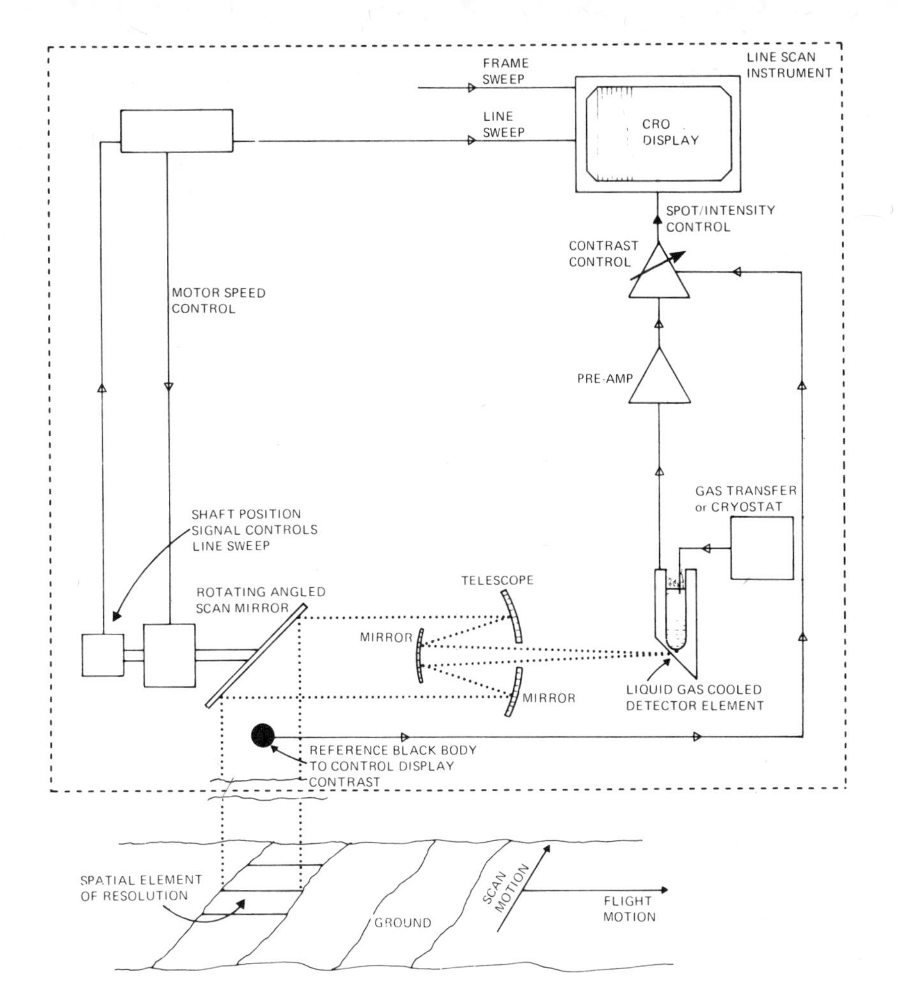

Fig. 6. Schematic of airborne thermal scanning apparatus used in remote sensing for civil and military purposes.

virtue of heat sensing being that the enemy cannot detect when they have been detected — as is the case with radar. The prime purpose was simply to detect the presence of the source but the trend moved to establishing thermal pictures of objects so that they might be identified, especially at night or in fog.

Thermal pictures are produced by scanning the detector across the scene in a systematic manner, as is done in television. It is necessary to scan the detector, for matrices of detectors are not yet available with sufficient detectivity and spatial resolution. Infra-red television camera tubes exist and they are used occasionally. One use is to plot a thermal map of a tool-bit whilst it is cutting in a lathe operation. Currently attempts are made to develop adequate arrays of detectors but in general, thermal imaging instruments still use mechanical scanning arrangements to sweep a single detector across a scene.

The scanning arrangement generally consists of a motor which drives a scanning optical element. This may be a rotating square, prism or flat with mirrored surfaces or it might be an oscillating mirror. This method — in effect — scans the detector across the scene, (Fig.6), for it is desirable to have the detector stationary especially when cooled. Synchronized to the drive, by electrical or mechanical means, is a moving light source which exposes a film or appears as a moving spot on a CRO screen. The brightness of the spot is controlled by the output of the detector which is decided in turn by the temperature of the element of scene being viewed. A line is thus reproduced on the film or screen which has temperature transformed into visible luminance level. In airborne units the second axis of the picture is provided by the flight movement — as the aircraft flies it produces a continuous strip thermal map of surface features below. In Figure 7 are scenes produced in this way. The power lines are quite clear: even their thermal shadows are to be seen.

Stationary thermal imaging devices have a second scanner, usually a large nodding mirror, that provides the frame fillup. As the frame rate is considerably slower than the line rate, the mirrors can be large. Their frequency of oscillation is low.

Military thermal scanners were the first to be developed and they are at least as sensitive as being able to detect the heat wakes where a ship passed many hours before, or where planes or vehicles stood on the ground. As the

Fig. 7. Thermal maps of countryside in California made with a Daedalus thermal scanner. Note especially the wind shadows and what could be thermal shadows of the power lines.

details were released from official secrecy, civil interests blossomed. Now under the name of remote sensing, (which also includes other radiation methods as well as IR and visible) countries are observing, to name just a few uses, the cloud formation from satellites, the mineral potential of previously poorly prospected areas, the movement of ground water, effluent discharges at sea, and thermal currents around power station cooler discharge channels. Recent publicity has been given to the Earth Resources Technology Satellite (ERTS). This was launched by NASA in 1973, and participating countries now receive data on the thermal emission of the land as seen as the satellite sweeps across, from 900 km up. The satellite's salient features are shown in Fig. 8. Differences over short and long periods will assist geological exploration, agricultural problems such as pest infestation and the control of pollution.

ACOUSTIC THERMOMETRY

There are some applications where the more generally accepted methods described so far are not appropriate

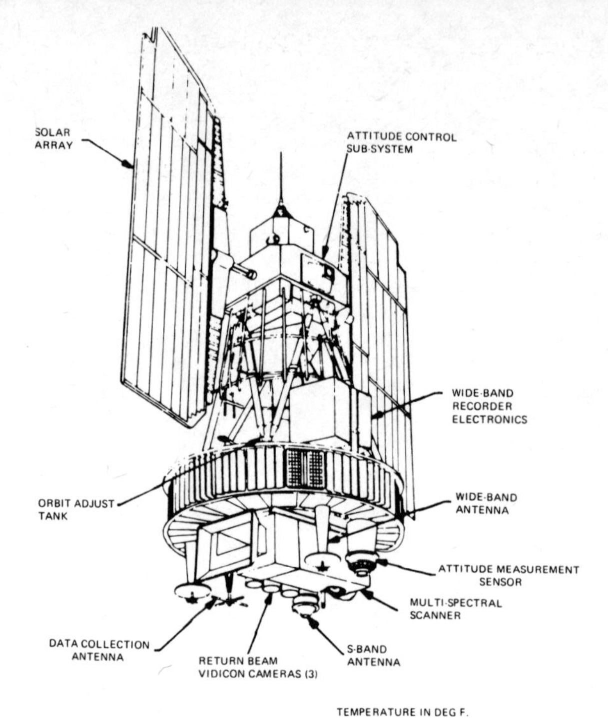

Fig. 8. The ERTS satellite currently orbiting the Earth to provide thermal emission data over long periods.

for one reason or another. Acoustic thermometers may fill the need, especially at the extremes of the temperature scale, for example, in gas plasmas (15 000°C) or in cryogenic systems near absolute zero.

The velocity of propagation of sound usually decreases as the temperature of solids and liquids rises – and conversely in gases. The velocities of sound propagation in various media are given in Fig. 9. The method then, is to determine the speed of sound and compute the temperature of the medium from this, knowing the velocity at a known temperature.

Two ways of doing this have been developed. The first way is where the sound wave is launched in the medium itself; the alternative is where the waves travel in a secondary material which is in thermal equilibrium with the medium. In either case there are two choices. A pulse can be transmitted, and the flight time measured; alternatively, the resonant frequency may be determined. Again, both methods require a sending transmitter and receiver although one may double for each if reflections occur. One pulse technique, known as the sing-around method, emits a send-pulse as one is received. The system then resonates giving a frequency-variable output with temperature change. Piezo-electric crystals are sometimes employed to couple to the cavity of interest. Magnetostriction is also used. Quartz crystals have already been mentioned in their role as temperature sensors – their operation is also acoustic. Thin wires have been used to sense temperatures in nuclear reactors (gas-filled cavities, discussed below, could be grouped as acoustic).

Other methods using sound that could be useful are to monitor the echo returned between dissimilar metals, for the reflection alters with temperature; to look for wave bending due to thermal gradients, and to sense when a solid surface becomes liquid and starts to reflect.

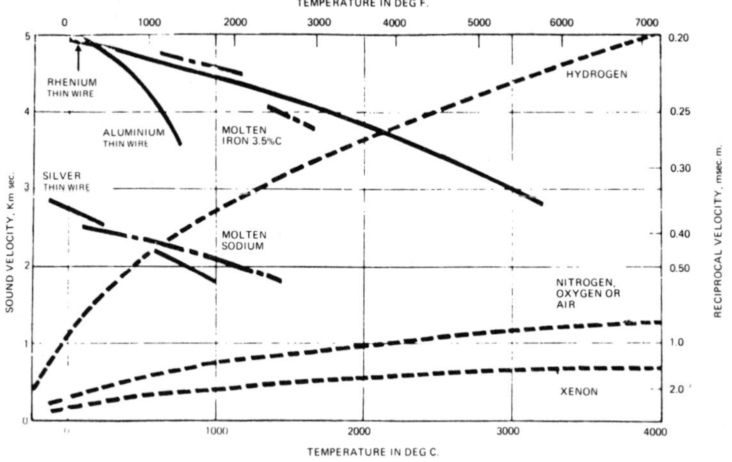

Fig. 9. Propagation velocities of sound in various selected media.

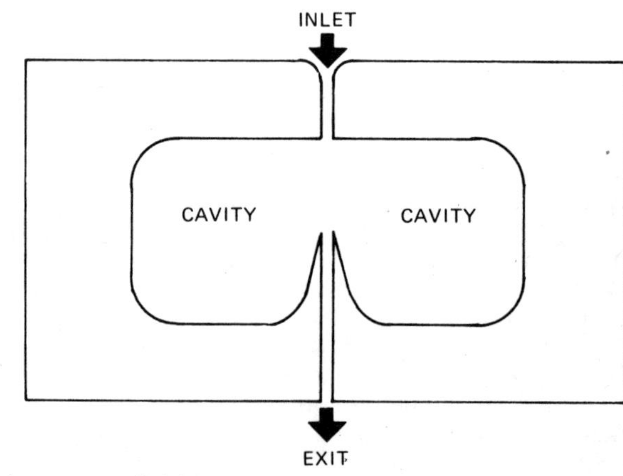

Fig. 10. Cross-section through a resonant fluidic temperature sensing cavity.

MISCELLANEOUS TEMPERATURE DETECTORS

Fluidic Sensors

A cavity filled with a gas oscillates at a frequency dependent upon the gas temperature. The upper limit of operation is currently decided by the sensor material. Hypersonic X15 aircraft used fluidic sensors to measure temperatures from 0°C to 2500°C. Gas flows in, past the cavity, shown in Fig.10, and is exhausted. The frequency of oscillation within the cavity depends upon the velocity of sound in the gas. For a given gas and size, the frequency varies as the square root of the Kelvin temperature of the gas. Typical output frequencies lie in the range 5 to 40 kHz. Response time is limited partially by the time taken to cycle the gas, and more so by the time for the complete cavity volume to change. Even so, rapid response is possible.

Eddy Currents

Temperature changes alter the conductivity of electrical conductors. By placing a sensing inductor coil near the test specimen it is possible to monitor temperature using the changes in impedance of the coil. This method is subject to many unwanted systematic errors, for example, changes in coil resistance with temperature and because of such errors the desired effect is very small. For example, only 0.05% change will occur in the coil for a degree change in an adjacent aluminium plate.

NON-ELECTRICAL TRANSDUCERS

A number of non-electrical output devices also exist to sense temperature. If a component, irradiated with the isotope Krypton 85, is raised in temperature, some of the isotope is initially released but then stops. To determine the temperature rise the part is heated until release just occurs again – and that is the temperature to which it was subjected in service.

Paints, pellets, crayons and liquid crystals (the cholesteric form) are available that change colour with temperature, some remaining, others returning when cooled.

FURTHER READING

Several of the references given in Chapter 6 are relevant to the radiation methods discussed in this Chapter. Acoustic methods, however, are not described in general texts as yet.

"Infrared Systems Engineering", R.D. Hudson, Wiley. 1969.

"Infrared Radiometry", J.R. Collins, Electronics World, Oct. 1967, pages 23-27 and 69.

"Using Infrared Thermometers effectively", H.L. Berman and M.R. Wank, Optical Spectra, July, 1969, 77-80.

"The Selection of a Biothermal Radiometer", D. Mitchell, C.H. Wyndham and T. Hodgson, J. Sci. Instrum., 1967, 44, 847-851.

"C.S.I.R.O. and the Australian Programme for the Earth Resources Technology Satellite (ERTS)", M.J. Duggin, C.S.I.R.O. Mineral Research Laboratories Investigation Report 95, July, 1972.

"Sound Ways to Measure Temperature", L.C. Lynnworth, Instrum. Technology, April, 1969, 17, 4.

"Ultrasonic Thermometry", S.S. Fain, L.C. Lynnworth and E.H. Carnevale, Inst. & Cont. System, Oct. 1969, 42, 107-110.

"Precision Radiometry", Vol. 14 of Advances in Geophysics". Academic Press, New York, 1970.

"Remote Sensing; Techniques for Environmental Analysis". Hamilton Publ. Co., 1974.

"Photometry and Radiometry for Engineers", Wiley, New York, 1974.

"An Infrared Pyrometer for the Measurement of Nylon Thread Line Temperatures", H. Bevan and R.E. Ricketts, J. Sci. Instrum, 1967, 44, 1048-1050.

"Temperature Measurement of Particulate Surfaces", J.H. Bach, P.J. Street and C.S. Twamley, Jnl. Phys. E: Sci. Instrum. 1970, 3, 281-286.

CHAPTER 8
MEASURING MOISTURE

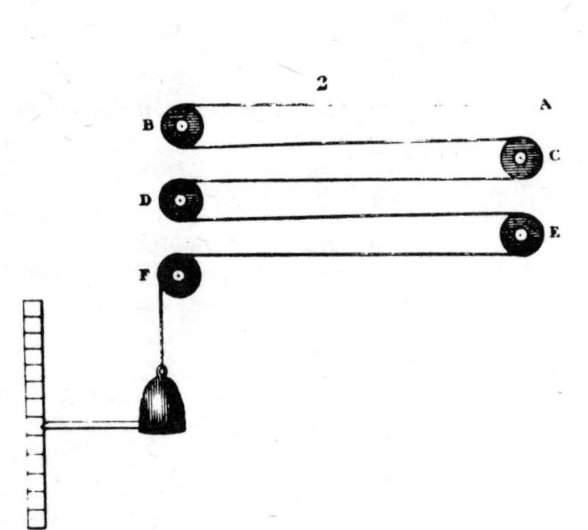

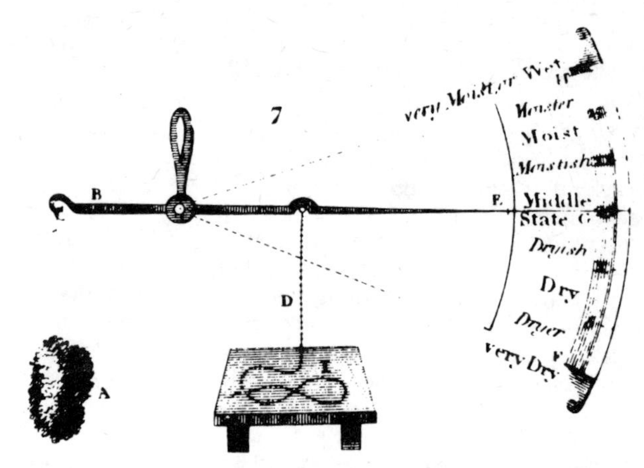

Instruments for the measurement of moisture in the air ... 1810 style.

KNOWING the water content of a substance is often of vital importance. In paper and board making, the moisture content of the pulp must be controlled; storage of valuable art objects and books demands humidity control of the environment; in the manufacture of synthetic textiles the air must be moist to prevent static electricity building up; personal comfort depends much upon the moisture present in the air.

In some areas of research it is necessary to monitor humidity in order that instruments are operated safely, for few are designed to work in saturated air. The laser interferometer method described in Chapter 2 requires correction for the humidity of the radiation path if accurate length measurements are to be obtained. In wheat milling, moisture content is important for it largely controls the amount of insufficiently ground grain that must be sold as reject material. The growth of plants depends upon the moisture content of the soil, plant and atmosphere.

In the study of radio propagation, it is necessary to determine rainfall and its droplet size distribution, for rain alters transmission significantly. Drops of rain impacting on a surface can cause erosion by impact and cavitation effects. In instances such as soil or structure erosion, the amount and size distribution of the drops is important. This Chapter deals with the techniques used to transduce water content data such as these into more convenient signals which are usually but not exclusively of electric form.

DEFINITIONS OF WATER CONTENT

Many definitions of water content exist, and it is necessary to have a basic understanding of these in order to comprehend the techniques employed to monitor it.

Moisture can occur mixed with other carrier gases, as found in the atmosphere, or it may be of interest when combined with solids to form a substance such as paper pulp.

In 1801, John Dalton, a British chemist, formulated Daltons law which states that the total pressure of a mixture of gases (or vapours) is equal to the sum of the pressures of each constituent gas if it occupied the same volume by itself. These individual pressure values are called the partial pressures. He also speculated that all gases would liquefy if the temperature were sufficiently low. Consequently from this, we can see that if a gas containing water vapour is cooled, there comes a point where water commences to liquefy or condense out. This temperature is the **dew point:** it occurs where the vapour is 100% saturated with water. The greater the water content, the sooner the water condenses as the temperature is lowered. Unless the temperature is raised, the gas will take in no more water in vapour form. The same concept can also be compared with the state change to ice (the solid phase of water) instead of liquid: this is known as the **frost point.** These two points each define the partial pressure of the water vapour but not in a single manner. Various institutions such as the Smithsonian Institution in the United States (see reading list) publish tables enabling the water vapour content to be assessed if the dew or frost points are known.

If the partial pressure is known, then the numerical value of any other definition of humidity can be obtained.

Other definitions commonly encountered are –

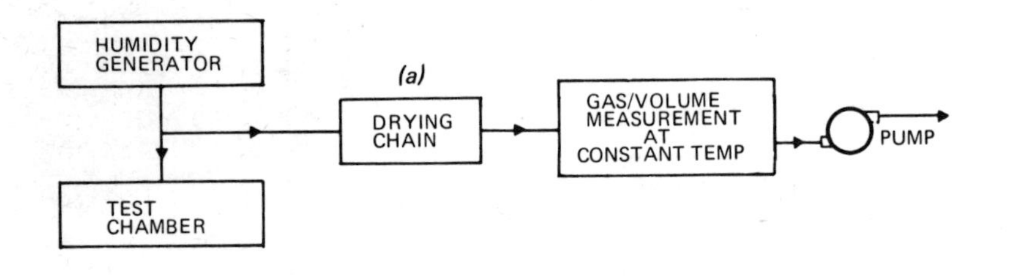

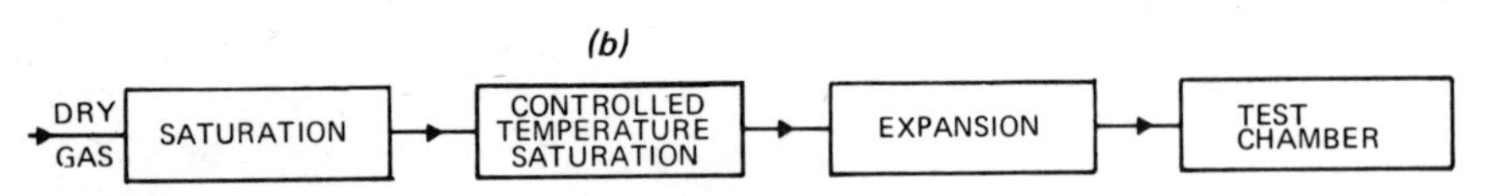

Fig. 1. Layout of the calibration standards for humidity.
a) gravimetric determination of moisture content (the absolute standard).
b) two pressure humidity generator (the transfer standard).

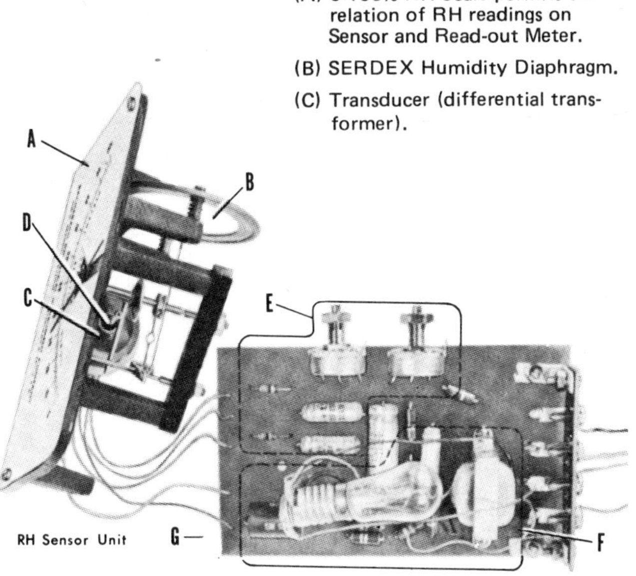

(A) 0-100% RH Scale permits correlation of RH readings on Sensor and Read-out Meter.

(B) SERDEX Humidity Diaphragm.

(C) Transducer (differential transformer).

(D) Transformer Core

(E) Differential Demodulator

(F) Temperature Stabilized Oscillator (approx. 4 kHz)

(G) Printed Circuit Board

Fig. 2. Operation of this SERDEX humidity sensor is based upon a hygroscopic animal membrane forming a diaphragm.

RELATIVE HUMIDITY, RH for short – this expresses the amount of water vapour present, compared with the maximum that could be at the temperature of interest. (The quantity of water vapour present would only apply for a stated temperature). RH is expressed as a percentage. For example, a dry day in the summer could be as low as 20% whereas when it is actually raining, it rises to 100%. The need for this relative unit occurs because many processes do not depend upon the absolute water content, but on the amount that could be absorbed or liberated from the air. RH is probably the most commonly used unit outside of process control areas.

PARTS PER MILLION, PPM – this expresses the water content by virtue of the weight of water, PPM_W or its volume, PPM_V, so it is either the ratio of the partial pressure of the water vapour to the total pressure, or else the PPM_V value multiplied by the ratio of the molecular weights of water to the other gas to yield the first value. Care is needed to define which is intended, for both units are dimensionless, and appear the same unless qualified with a (by weight) or (by volume) statement.

WET BULB TEMPERATURE (no accepted abbreviation exists) – if a thermometer has its sensing area wetted with water (usually with a saturated wick) and air is passed rapidly over it, the thermometer reads a value less than that of an identical dry thermometer by an amount depending upon the relative humidity. If the air is 100% saturated, no more moisture can be taken up so the bulb is not cooled at all. (The same reason is why evaporative air coolers do not give as much cooling in humid weather). This concept is used in the wet-and-dry bulb hygrometer.

MIXING RATIO – the ratio of weight of water vapour to dry carrier gas.

POUNDS/KILOGRAMS PER HOUR – this expresses the absolute amount of water vapour supplied per hour. For example, heat treatment of metals requires knowledge of the water content in the furnace as this controls the carburizing process reaction rate.

RELATIVE EQUILIBRIUM MOISTURE (rem or em) – in the paper industry, it is the ability of the fibres to lose or absorb water (the sorption process) that decides the shrinkage, tearability, etc. Equilibrium will eventually occur between the air humidity and the paper content. To make it clear that it is the paper moisture content that is stated, em is quoted. Hence a lower em than RH means the paper takes in moisture.

LIMITATIONS OF DEW POINT MEASUREMENT

Not all moisture measurements make direct use of the dew point phenomenon but it is instructive to consider the limitations of the process for the effects are present in most procedures.

THE KELVIN EFFECT – In 1870, Lord Kelvin arrived at the conclusion that the vapour pressure over a concave liquid surface is less than that over a plane surface of the same material. Water condensing on a surface forms droplets which produce a curved interface surface with the surrounding vapour. It has not been an easy matter to prove Kelvin's theory, for the effect is small, but convincing electron microscope studies of evaporating lead, carried out at Imperial College in London, have shown it to be true for lead and gold. The Cambridge Systems Company of Massachusetts have estimated the depression in dew point temperature due to 30 μm dropsize condensation as 0.005 K. Few people would find the Kelvin effect error a problem.

THE RAOULT EFFECT – In 1887 Raoult produced a law governing the

Fig. 3. The Yellow Springs Instrument Co. aspirated psychrometer. The righthand view shows the unit packed for transportation. Thermistors are used as temperature sensors.

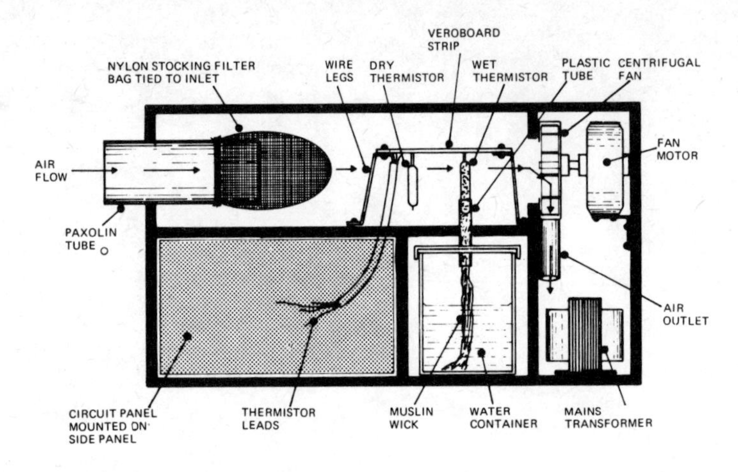

Fig. 4. Inside layout of an aspirated wet-and-dry bulb hygrometer which uses computing circuitry to produce a RH output.

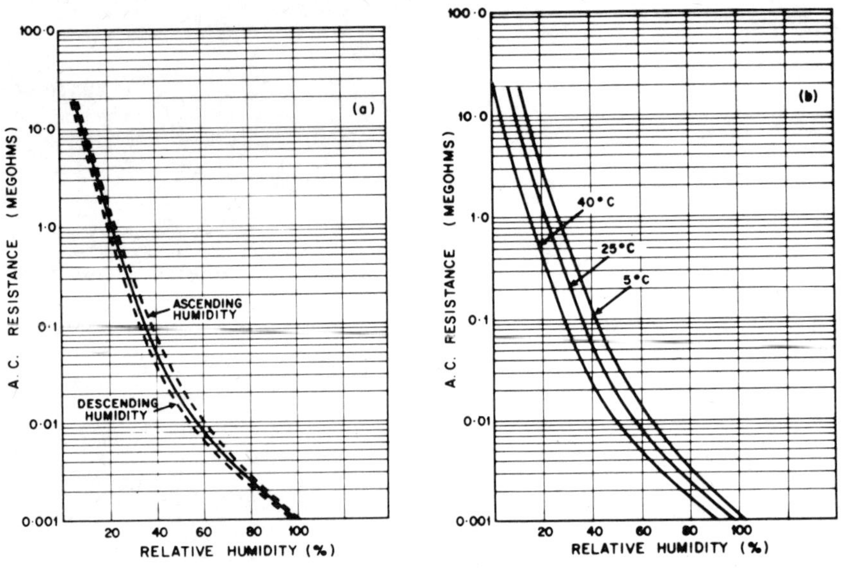

Fig. 5. Performance curves for the Warren Components Corporation humidity sensing element.
a) these curves show the lag error depending on the direction of approach to a value.
b) here are shown curves of resistance variation with humidity. Note the shift due to temperature and the slope away from a truly logarithmic response at high RH.

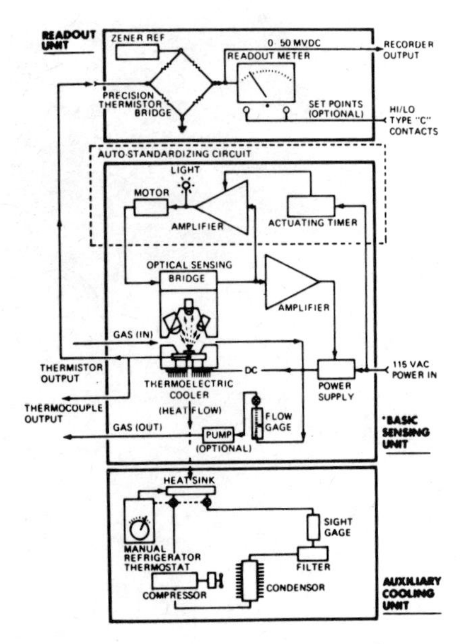

Fig. 6. System configuration of the Cambridge Systems dew point gas analyser. (The detector is in the centre of the system).

Fig. 7. Exposed view of the sensor used in the system of Figure 6.

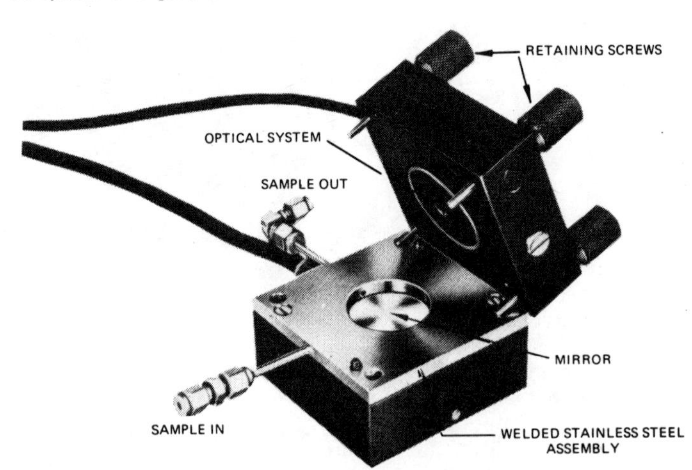

effects of impurities on vapour pressures. If contaminants exist upon the dew forming surface, the vapour pressure is decreased. It has been estimated that this produces an error equal but opposite in sign to the Kelvin effect error if the same surface with its 30 μm drops is contaminated with a 10 molecule thick salt layer. These errors are random so they will not necessarily cancel. The Raoult error is reduced if there is more water on the surface diluting the impurity. As we shall see later, this heavy dew operation can be achieved by simple adjustment in dewpoint measuring methods so is preferred in exacting applications.

THERMAL MEASUREMENT ERRORS — the determination of dew point involves temperature measurement so this also needs to be of adequate accuracy. The two previous chapters dealt with temperature measurement so it is only necessary to reiterate that errors due to sensor calibration, heat loss and gain due to mounting thermal conductivities need consideration if accurate results are to be obtained. Generally, thermal measurements limit the precision of dew point devices.

STANDARDS AND CALIBRATION

The standard used to calibrate moisture determining instruments is called a gravimetric hygrometer. This uses a procedure, shown in Fig. 1a, whereby the water vapour is absorbed in chemicals, leaving only the carrier gas — which is weighed with great precision. The process has greater accuracy than other methods (as it should have, being the standard) but is bulky and time consuming requiring many hours to make a determination.

A second, less accurate device, known as a two pressure humidity generator, (see Fig. 1b), is used as a transfer standard (its value is set in relation to the standard hygrometric method and can be used to test many instruments before needing recalibration). In this method a dry gas is saturated and then passed into a superior controlled system where the saturation is ensured by using temperature control of the vapour. It is then passed into an expanding chamber where the device to be tested is housed. The degree of expansion decides the vapour pressure of the water. In field or factory use, neither of these is satisfactory due to size, cost and time factors. Instead a substandard, preferably a dew point device, is used which, (if warranted) is calibrated by a standards laboratory. The wet-dry bulb method is sub-standard (in order of standards hierarchy) to the dew point methods, the tables used with it having been derived from dew point data.

A simple method to produce a test environment is to inject a known amount of water into a known volume evacuated chamber. Dry inert gas is then introduced to provide the pressure needed. This idea has been used at the von Karman Gas Dynamics Facility in the United States to provide a calibration accurate to a dew point error of ± 0.8 K. This is not as accurate as the ultimate standard methods but does suffice to check and intercompare most daily-used techniques.

Gravimetric equipment is capable of dew point determinations to about 0.01 K, the two pressure generator to 0.06 K and the dew point method to 0.25 K. Other methods are generally inferior, but each must be considered, for the specific application may render accurate methods inaccurate due to peculiar factors.

MEASUREMENT OF HUMIDITY

With this background, it is now possible to discuss actual techniques.

HYGROSCOPIC MEMBRANES AND HAIRS

The least scientific but easiest principle to employ makes use of the fact that some organic materials alter dimension with changing moisture content. Human hair, for instance, is used in laboratory clockwork driven humidity recorders. It extends some 3% for the change from zero to maximum water content. Several commercial instruments use the movement produced to drive a microdisplacement transducer thereby providing an electrical signal — one unit, shown in Fig. 2, uses an inductive solenoidal transducer; another, a strain gauge placed on a driven mechanical member of the linkage operating the indicator pointer. Such devices could be expected to work to a few percent RH accuracy only and it is recommended that they be used only in the range 15–90% RH and in the temperature range 1–40°C. Their operation is prone to slip-stick effects and somewhat erratic response is experienced. They also need cross calibration to a more fundamental method.

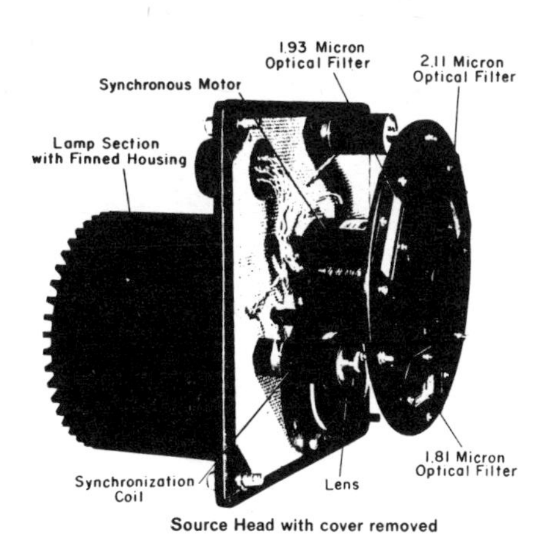

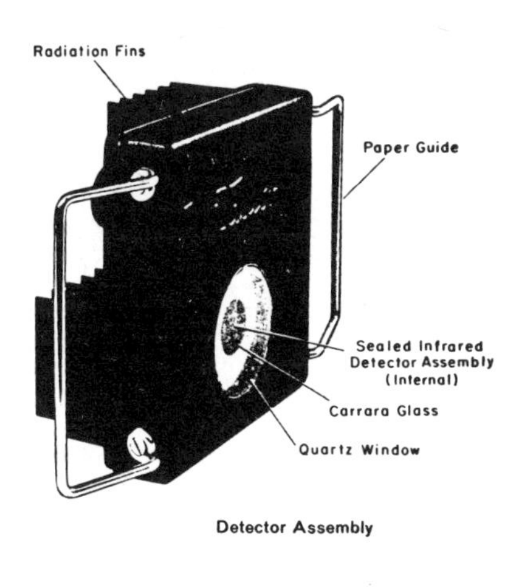

Fig. 8. Source head and detector assembly of the Taylor infra-red moisture gauge.

WET-AND-DRY BULB PROCEDURES

The simplest wet-and-dry bulb hygrometer consists of two mercury thermometers in a frame, with the bulb of one covered by a wet muslin sleeve. The evaporative effect is at the mercy of draughts so this static procedure is not very accurate. Dew point temperatures are not achieved on the wet bulb. Rough estimates can be made of the dew point using Glaishers factor (which varies with the temperature) but for best work standard tables and charts are used.

A better arrangement is to cause the atmosphere to pass over the wick at high speed. As the surface velocity is increased, the cooling effect falls off as the air velocity is increased, and at around 30 m/min there is little to be gained by it going faster. The obvious development step, therefore, was to whirl the two thermometers. Devices made this way are called whirling or sling psychrometers. (The grand name is applied without real distinction to the term hygrometer). One maker supplies a unit with a built-in slide rule scale that gives RH directly from the wet and dry readings taken after three minutes of rotation. In general, it is necessary to use the charts. Where electrical signals are needed or where the delicate glass thermometers need to be omitted, other methods are used.

As it is the relative wind speed that matters, it is easier in wet-and-dry instruments to aspirate the two thermometers by driving the air over them using a fan. The Assman psychrometer uses a clockwork or electrically driven fan to do this. This form of instrument is amenable to automation by replacing the mercury thermometers with proportional readout electrical sensors. Commercial units are available that aspirate thermistor sensors, (see Fig. 3), reading wet bulb depression directly.

It is still necessary, however, to resort to tables to obtain values of RH and other units with these. In 1969, 'Wireless World' reported a design that went one stage further. An inside view of the device is given in Fig. 4. The thermistor resistance variations were firstly linearized and then combined in an operational amplifier arrangement to yield a direct linear scale of RH to ± 5% accuracy. It has several short-comings, namely not being useful near freezing point temperatures (as is the case with all wick-type devices) and at low RH values. It does, however, show that a direct reading aspirated wet-and-dry bulb method can yield RH values without the need to use tables making it useful in automatic control or recording applications.

RESISTIVITY AND CAPACITIVE METHODS

The resistance between electrodes connected to a moist substance such as paper or soil is a measure of the moisture content. On paper making machines, wiping fingers have been used; in soil, electrodes embedded in a plaster block have been suggested. The variation of resistance with moisture content is roughly logarithmic but the actual value depends much upon contaminants and salts in solution that produce electrolyte. The use of ac measuring techniques is superior as this eliminates the electrolytic effect.

Evaporated grid sensors (gold fingers interleaved over an inert substrate) are made in which resistivity varies. Akin to this is the Warren sensor which consists of a special plastic backing having a humidity sensitive conducting

Fig. 9. Microwave moisture meter (Skandinaviska Processinstrument AB, Sweden) mounted in line for monitoring of moisture of materials flowing in tubes.

plastic grid on the surface. Again the response is logarithmic and a small temperature coefficient and hysteresis effect exists — as Fig. 5 shows. The Dunmore sensor (originated in the 1930's) is made as two wire spirals forming electrodes on a former which is coated with lithium chloride.

Changes in capacitance also can be utilised. Simple sensors use plates separated by the medium of interest to form a capacitor. Differential arrangements assist in reducing errors. For air humidity sensing, a sensor has been made from porous anodized aluminium strip, coated each side with gold layer electrodes. This acts as an aluminium oxide capacitor in which moisture diffuses into the pores.

Although a simple matter to measure the resistance changes and display them on a calibrated non-linear scale meter, it does not yield dew point nor give a linear scale. The overall accuracy of an instrument using these detectors can be improved if the humidity sensor is used only as an error detector of dew point.

DEW POINT DETECTORS

The most obvious way to detect the appearance (or disappearance) of moisture is optically to monitor the reflectance of a mirror surface. Several devices do just this. Simple instruments are manually observed. A mercury thermometer is placed in a heat conducting block having a mirror surface. Cooling is applied uniformly with ice or ether to the block as the air is aspirated across the mirror. The temperature, when mist occurs, is the dew point. (In practice, a rough run is made first, with a second slower rate taken as the system heats up to lose the condensate.) Improved designs are still being reported in the scientific literature. They are inexpensive and accurate to 1 K.

Automation has been achieved by viewing the reflected light with a photo resistor. One system using this is shown in Figs. 6 and 7. A stainless steel mirror is cooled by a thermoelectric Peltier cell until reflectance drops as the condensate forms. It is then temperature controlled to track a constant reflectance situation. The block temperature is held at the dew point of the gas passing over it. Temperature is transduced by thermistor or thermocouple in the unit shown. By altering the level of the reflectance signal, it is possible to operate with a heavy dew reducing the Roault errors. A precision instrument is actually more sophisticated in its peripherals, as can be seen in Fig. 6.

Resistance and capacitance sensors can also detect the presence or not of moisture, and many dew point meters use them instead of the optical method, for there is a sharp transition when the condensate is boiled off as the element is heated.

Where size is vital, the Spanner method (after D.C. Spanner in 1951) can be used. This uses two thermocouple junctions, one dry, the other in the active environment. The generated voltage is read for the dry unit. It is then used as a Peltier junction, cooling it for a timed interval. Water condenses upon it. The circuit is then switched back to measuring: the voltage measured is related to dew point. The principle has been used by many people since 1951. The procedure has been largely automated. The low cost of the sensors makes the method very applicable in field studies of soil-plant-water relationships where many tens of measurements are needed. A single cyclic switching and interrogating unit is used to operate the sensors in turn.

HYDROLYSIS

A thin film of phosphorous pentoxide, held between noble metal electrodes, absorbs the moisture of the sampled gas as it flows through. A dc voltage across the electrodes breaks the water down into hydrogen and oxygen. The electrolysis current flowing is a measure of moisture content and will cover ranges from 0 — 100 to 0 — 10 000 parts per million with about 5% accuracy of the full scale setting chosen.

ENERGY ABSORPTION METHODS

Gas molecules absorb electromagnetic radiation in a selective manner. Water vapour attenuates energy in the region of 3–6 μm wavelength (infra red). Microwave frequencies in the region of 100 to 4000 MHz are also strongly attenuated by moisture vapour.

In both cases, the same idea is applied. A source of the radiation (a tungsten lamp for IR work or solid-state low-power microwave sources) is viewed by a sensor through the gas. Measuring the difference between the ideal signal and the absorption signal enables the loss to be determined, and hence the moisture content. In the IR instrument, the difference may be obtained by rotating a filter wheel so that the detector sees the source through the sampling cell — first with a bandpass optical filter and then with a selective absorption band filter. Another method, shown in Figure 8, sequentially applies three (two have been used also) IR sources (the same tungsten-halogen broadband lamp but with different narrow band-pass filters to select wavelengths needed) to radiate through the gas stream onto a lead sulphide detector. This provides three levels of signal which can be combined and processed to give the moisture content in the required units.

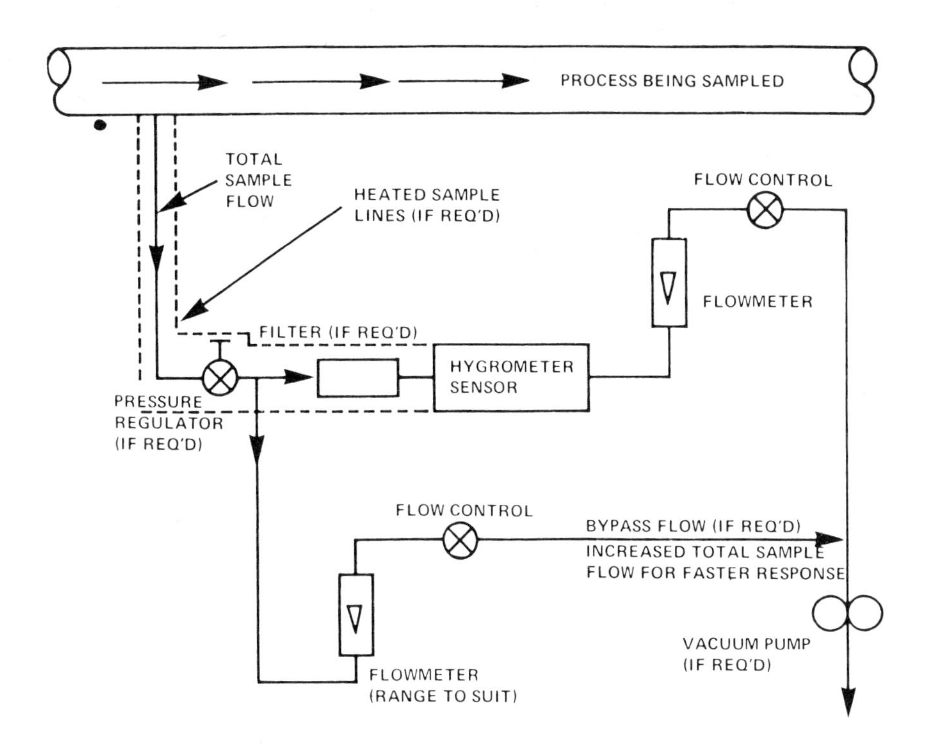

Fig. 10. Components and layout of a moisture sampling system.

Most microwave devices operate by monitoring the amplitude loss as it is easier to measure attenuation, and experience in such measurements is commonplace. The use of phase shift effects instead is relatively new, but at the Royal Institute of Technology in Stockholm whose unit is shown in Fig. 9, a number of these (which measure dielectric effects) have been built proving their feasibility.

INSTALLATION OF MOISTURE MEASURING DEVICES

Many instruments are operated continuously to provide moisture measurement. Where the through volume is large a sampling line is bled off to feed the detector. This line may introduce errors by contaminating the input gas with leaks or chemical reaction or more likely by condensing moisture out of the gas. The line temperature must never drop below the dew point so heated lines may be needed. Fig. 10 is a schematic layout of a sampling system recommended by Cambridge Systems. Filters may be needed to remove solid contaminants but usually dew point meters include these. Flow meters are vital in order to know the amount of gas sampled. Teflon or stainless steel tubing are used for they are non-hygroscopic.

If the hygrometer is multiplexed between a number of lines, problems can occur due to the time taken for the materials of the lines to lose the absorbed moisture. The lower the dew point value, the longer it takes to purge the lines so switching between widely differing samples will have a restricted response. For example, a poor line (nylon for instance) takes several hours to re-establish equilibrium. Lines are usually purged clean with dry air or nitrogen before use. Freon 114 is an excellent solvent gas to use.

RAINFALL

It has become necessary to use higher frequencies (10 – 30 G Hz) in communication links to gain more bandwidth. Unfortunately, at these wavelengths, rain can seriously attenuate signals, so the major laboratories (Bell Telephone and the British Post Office, for example) have research programmes going to investigate this. One of their first problems was to produce continuously recording rain gauges, for the manually-read weather station rain gauge only gives values integrated over long time intervals.

The Bell system makes use of the fact that rainwater has a high dielectric constant. A collecting area gathers rain, funnelling it into a narrow run-off channel. Inserted in the channel are electrodes which are normally separated by air. When rain spills down the chute, the dielectric constant between the plates changes (in proportion to the amount flowing) and this is used to alter the capacitance of an R-C oscillator. The output appears as a frequency deviating signal. The Bell company placed 96 of these units over a 50 square mile area and telemetered the data back to a central recording unit. The data were processed to produce isometric plots of the rainfall, as shown in Fig. 11. It was then correlated with the signal loss of microwave transmissions.

The British groups (Post Office Research Department and the Radio and Space Research Station) recently reported another technique. The collected rain-water is fed, (see Fig. 12), to a 3 mm internal-diameter tube where it drips through as constant size (but varying rate) drops. The falling drops are detected with a simple photo transistor light interruption sensor. A digital unit telemetry link is used to send an 8 bit binary code of the drop rate to the processing centre.

RAIN DROP SIZE DISTRIBUTION

In many instances where rain fall is measured, it is necessary to know more than just the amount of water precipitated. For a number of reasons (erosion being the most general need, for the droplets can damage a surface – soil for instance is eroded by impact and washing) the research worker needs to obtain data on the nature of rainfall. Drop sizing and distribution (rain spectra) is a recurring measurement problem. As early as 1904 a simple method of flour balling was used (flour encases the drop as it falls into it). High speed photographic methods have been reported regularly. The advent of electronics enabled drops to be monitored faster and more accurately. The list of methods is extensive.

A recently reported development by staff of the I.I.T. Research Institute in Chicago, uses a television scanning arrangement to count and size the droplets as they pass a viewing area. The drops are virtually frozen by using a rapid flash of a xenon lamp to illuminate them in front of a vidicon tube. During this period, the scene is scanned and line data stored. The video signal is then processed to decide the quantity of drops and their size distribution and these data are printed out using a number of channels. To test the instrument routinely, slides which have drop cross-sections on them, can be inserted in the optical space.

For best performance assessment,

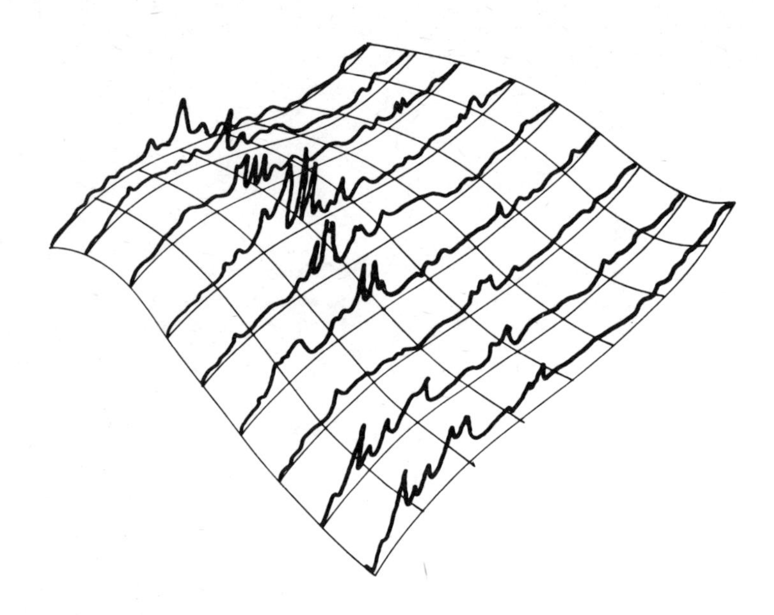

Fig. 11. Three dimension representation of rainfall distribution measured by Bell Telephone laboratories

water sprays and falling glass beads were used whose distribution had been proven by independent methods. In one method a belt of sensitized paper was moved past a test slit. Drop imprints on the paper through the slit were made permanent by exposure to ammonia fumes. The method is similar to many sizing systems now in use (blood cell and leaf area come to mind) but here the image is dynamic and splashing drops on the terminal boxes of the equipment lead to errors.

Finally, a brief comment on ice measurement. On aircraft superstructure and North Sea trawlers, ice builds up impairing the performance; on the former it would lead to trouble so de-icing units, which heat or crack the ice away, are standard equipment. Generally, the pilot decides when to de-ice by subjective factors such as heaviness of control. Approval has recently been given to a small device that electrically indicates the amount of ice build-up. A probe about 25 mm across is fixed in the air-stream to simulate the worst point on the airframe. The probe is an oscillator in which its mechanical mass forms part of a tuned circuit. As ice builds up, the resonant frequency changes providing a continuously varying signal of build-up.

FURTHER READING

HYGROMETRIC AND PSYCHROMETRIC TABLES:
"Smithsonian Meteorological Tables", 6th ed., R.J. List, Publication 4014 Smithsonian Institute, Washington, D.C., 1963.
"Hygrometric Tables", H.M. Stationery Office, 1964.
"Psychrometric Tables (Marvin)" Published in U.S. available C.F. Cassella, Regent House, Fitzroy Square, London.

DEFINITIONS AND TERMINOLOGY:
Numerous working data pamphlets by Cambridge Systems Inc., Newton, Massachusetts 02158, U.S.A.
"Humidity and Moisture: Measurement and Control in Science & Industry" Reinhold Publishing Corp, N.Y., 1965, Four volumes.

CALIBRATION:
"Humidity Calibration Techniques", K.M. Cole and J.A. Reger, Instrum. & Control Systems, January, 1970, 77–82.

METHODS IN INDUSTRY:
Many papers appear in issues of "Measurement and Control" journal.

Fig. 12. The R.S.R.S. rain gauge.

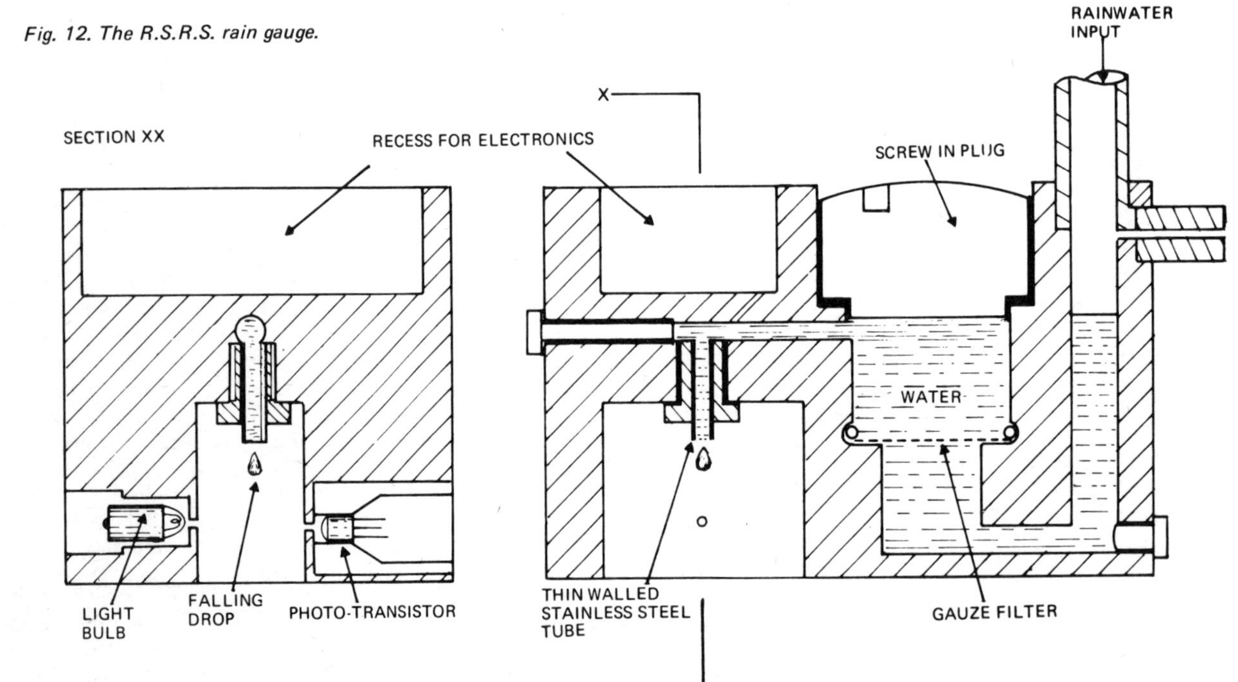

CHAPTER 9

FLOW

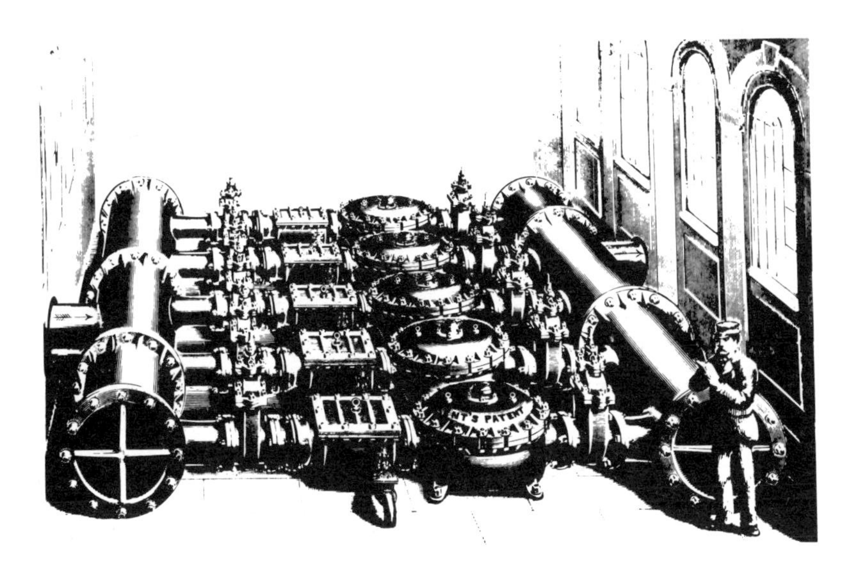

Largest world installation of positive-flow water meters in 1895... Corporation of Oldham, England. (George Kent).

Fluids in motion range from low density gases, through liquids to slurries, pulps and particulate substances such as wheat or sand. And they often require to be monitored to provide information on the rate at which a volume of fluid is passing a given area, or the rate at which the mass of the fluid is moving, or to monitor the total amount that has passed in a given time-interval. Examples are: pulp flow in paper making, blood flow in medicine, power station cooling-water rate, air speeds in weather forecasting, the flow rate of highly reactive liquid sodium in nuclear reactors. Flow rate measurement is also often required for delivery of exactly metered quantities of a substance.

As with all sensors, no universally applicable device or principle suffices for all the various needs. Many principles are invoked, in a variety of ways, in order to provide a satisfactory measurement from both performance and cost points of view.

FLOW CHARACTERISTICS

In general, most fluids needing measurement are conveyed in enclosed pipes, although in some instances open channels are used. It is, therefore, useful to know something about fluid transmission in pipes. Assuming the fluid is incompressible – water and oil come close to the ideal – then the flow rate is a measure of the volume (volumetric flow rate), or mass (gravimetric flow rate) passing a point in a given time, depending on which is of interest. The greater the velocity, the greater the volume (or mass) passing. However, a strict correspondence does not exist, and the relationship depends upon how the flow takes place in the conduit, which leads us to the two main types of flow that can exist in a pipe.

If the particles of the fluid flow in a smooth streamline manner – imagine the flow as numerous ultra thin layers slipping over each other with greatest velocity in the centre – then this is laminar flow. Empirical observation plus dimensional analysis of the

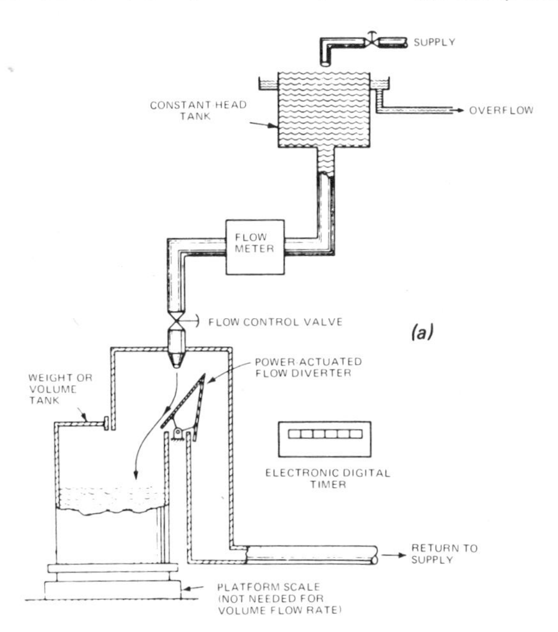

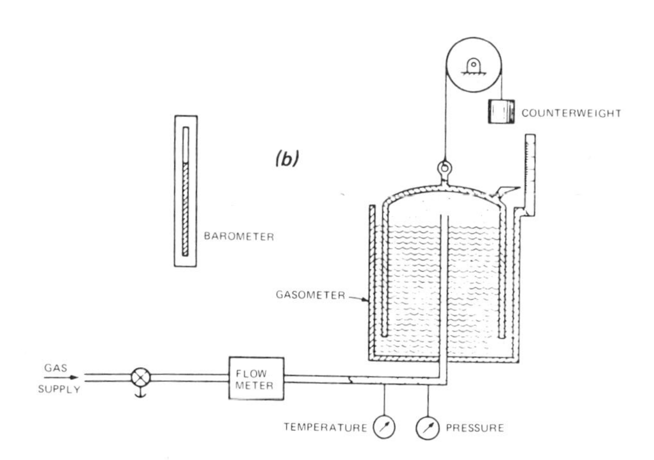

Fig. 1(a). Calibration of liquid flow meters.
(b). Calibration of gaseous flow meters.

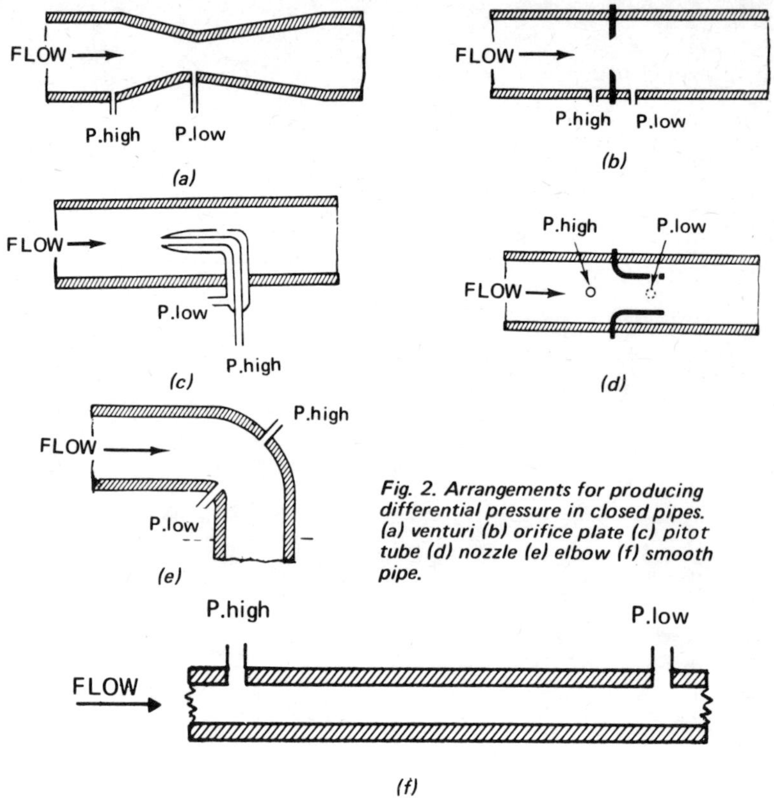

relevant equations of fluid mechanics have yielded a very valuable characteristic number that depends upon the pipe diameter, fluid density and viscosity, and the flow rate. This Reynolds number (after O. Reynolds, 1883) will be around 2500 or less if the flow is laminar. We shall see later that correct flow measurement often requires that the flow through the transducer is laminar, so a section of pipe or straighteners are added to steady the flow before it enters the transducer.

If the flow is laminar, the velocity of the fluid particles will vary across the pipe section, the layer being stationary near the pipe-wall and fastest at the centre of the pipe section. The velocity gradient depends on the fluid and the pipe, so the average flow rate could be quite different from that indicated by a flow meter.

The other distinct flow type is called turbulent because the velocity at any point in the flow is random and no streamline flow exists. A Reynolds number greater than 4000 usually indicates turbulent flow. In between the two numbers, the flow depends upon the pipe-work system used. Obviously, rough edges, surfaces and steps in the pipe will cause turbulence for some distance down stream. Flow meters are therefore usually installed away from elbows, joints and valves.

To add to the problems, flows are often pulsating due, perhaps, to the use of a gear or piston pump. In recent years considerable research effort has been devoted to the study of pulsating flows. If possible, the rule is to circumvent pulsation problems by smoothing the flow with a storage tank, (which acts as an integrator), or by mounting the meter some distance away from the source of pulsation. Some flow meters operate without significant errors in pulsating conditions, but not all.

Flow, therefore, is a challenging variable to measure. Extreme care is needed in the selection of the method, and its application. A good understanding of basic fluid mechanics is essential.

Other factors that must be considered are that the components of the transducer will not be corroded by erosive chemical or cavitation forces. They must also be able to withstand the temperatures involved.

The majority of flow meters are designed to operate with a specified flow direction only, their principle will not work correctly the other way.

FLOW → P.high P.low (a)

FLOW → P.high P.low (b)

FLOW → P.low P.high (c)

P.high P.low FLOW → (d)

FLOW → P.high P.low (e)

P.high P.low FLOW → (f)

Fig. 2. Arrangements for producing differential pressure in closed pipes. (a) venturi (b) orifice plate (c) pitot tube (d) nozzle (e) elbow (f) smooth pipe.

CALIBRATION AND TESTING

In theory, flow (being mass or volume passing a given area in a given time) can be defined in terms of the fundamental mass, length and time units. The only satisfactory basic standard procedure is to pass steady flow through the device, collecting the fluid in a suitable weighing or volume measuring enclosure and measuring the time of flow. Schematics of gravimetric and volumetric test setups are shown in Fig. 1. Flowmeters so calibrated may then be used as substandards (the absolute methods are time consuming) to calibrate other flow meters in continuously flowing closed-circuits. An essential requirement with absolute methods is that the pressures to and from the transducer are maintained constant to ensure an even flow rate. Closed-circuit systems are used but pressure control is also needed. It is not always necessary to measure the entire flow in the pipe — shunts are often used to bypass a known percentage of the main flow through a parallel mounted, smaller, flow-meter circuit.

Fig. 3. Solid-state mass flow and density computer for use with differential pressure devices.

Having dealt with this background, we can now examine specific designs.

Fig. 4. Cross-section of a simple variable-area rotameter.

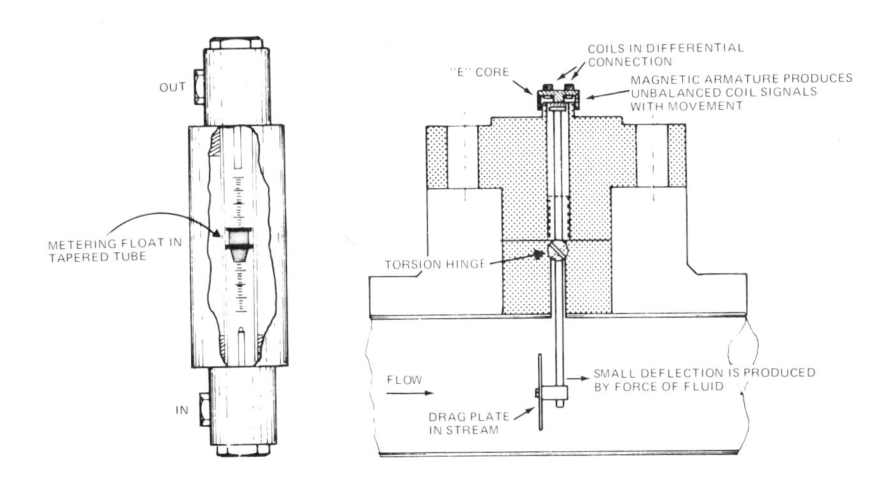

Fig. 5. Schematic of a drag-plate flow-meter having fast response, made at the National Engineering Laboratory.

DIFFERENTIAL PRESSURE (d.p.) SENSING OF FLOW

In the middle 1700's, Bernoulli studied the ideal case for steady flow of an incompressible fluid in a frictionless pipe. He was able to relate the variables of a flow between two points – viz, pressure difference between the two points, velocity of flow and difference in pressure head – using a general energy-balance equation that now bears his name. If flow is restricted by narrowing down a short section of pipe, it can be shown from this equation, and from consideration of continuity of flow, (the mass leaving the restriction equals that entering it) that there will be a pressure difference between a point upstream and a point in the restriction. A cross-section of such a venturi device is shown in Fig. 2a. (Carburettors and spray guns use this principle to draw vapour into the air flow passing through the venturi). Flow rate, therefore, can be transduced into an intermediate secondary variable, pressure difference, which can be monitored with pressure transducers. Flow, however, is proportional to the square root of this pressure difference and linearization is needed. Pulsating flows are not indicated correctly, due to this non-linearity.

A simple way to invoke the same situation is to insert a plate, having a small hole in the centre, in the flow stream. These are called orifice plates (Fig. 2b). Again flow velocity is dependent on (pressure difference)$^{1/2}$. The actual relationship depends critically upon the hole diameter, its profile and the fluid constants of viscosity and density. Every combination has a slightly different discharge coefficient. Standards have been established to ensure accurately related flows with the measured pressure differences. Orifice plates are available commercially. Other devices using the same concept in different ways are the Pitot tube and nozzles (shown in Fig. 2c and 2d).

A pressure difference is also produced between the outside and inside of a bend in the pipe work, (Fig. 2e), due to centrifugal force, and this is often used as a metering method. Again certain criteria must be adhered to, especially the use of a 'calming section' preceding the bend. Individual calibration is necessary.

Yet another pressure difference device makes use of the pressure drop developed along an even-bore pipe by fluid friction effects (Fig. 2f). Although providing a smooth bore to flow, the resistance needed to develop enough pressure difference may be a disadvantage. To shorten the length whilst retaining smooth flow, one manufacturer offers a pipe filled with glass spheres for the same purpose.

The basic low cost, reasonable reliability and installation ease, made d.p. methods popular. Many instrument companies offer equipment that linearizes the d.p. signal to indicate flow or mass on a linear scale. A solid-state equipment is shown in Fig. 3.

Pipe sizes from hair size to many metres in diameter can be instrumented this way.

Obvious disadvantages of such methods are the need to maintain the restricted area free of debris and solid contaminants. The method is not used for highly viscous or particulate substances as a general rule.

DISPLACEMENT DEVICES

An object placed in the fluid stream experiences a force attempting to move it along. Many devices make use of this fact to provide an intermediate stage by which a linear or rotary displacement sensor is actuated.

The simplest meter uses a spring-loaded horizontal or vertically suspended object in a tube such that movement of the object alters the area through which the fluid is restricted. These variable area meters, in fact, produce a varying size orifice which maintains a constant pressure drop – in contrast to the constant area of the differential pressure methods. Inexpensive units visually indicate flow as the position of the float against an engraved scale (as shown in Fig. 4). By suitable design, the movement can be made linear with flow rate. They also cover a wider range of flows (10:1 is possible) than differential pressure methods. To obtain an electrical signal, the movement is measured with displacement transducers such as the inductive sensors or potentiometers.

Another displacement type is the turbine meter. If accurate metering of valuable fluids is needed, the blades are designed with seals that slide to ensure minimum bypass of fluid past the blades. (Often the pump is designed to act as the flow meter. The disadvantage then is the need to provide energy to move the rotor, and this resistance to flow may be intolerable).

Where flow is established and reasonably steady, the blades need not be close fitting as some slip is allowed. The essential features of a turbine flow meter is a freely spinning turbine that couples closely to the flow, a flow straightener preceding it to avoid errors due to already rotating flows, and a non-contacting sensor to detect turbine rotation. Most flow meters use magnetic detection. A small magnet is inserted in the blade and the external sensing coil produces a pulse for each pole of the magnet passing it. Other sensors utilize capacitance/resistance changes; several designs also operate an indicating meter with a mechanical link.

Turbines can provide a linear output but because the indicated flow depends upon the (diameter)^2 they are made as a complete unit calibrated for

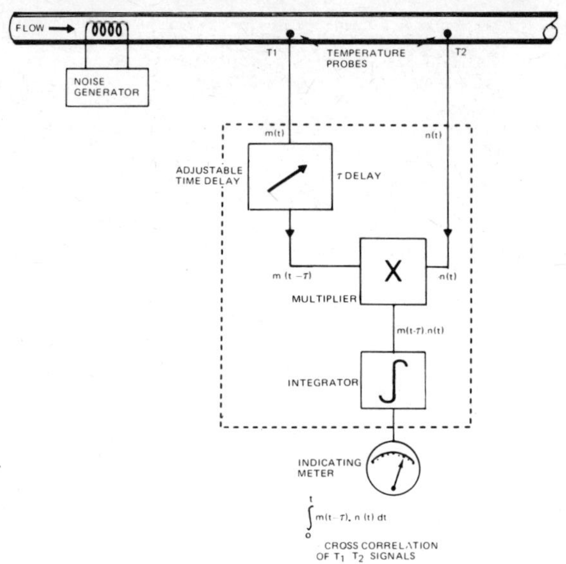

Fig. 6. Principle components of a cross-correlator thermal flow-meter.

the bore used. The output is a frequency variable pulse signal. This aids reliability and ensures compatibility with digital equipment. Such devices have been developed to measure liquids and gases, covering a range of flows of 15:1. Using different size turbines, gas flows from 1 m^3/hr to 50 000 m^3/hr can be measured.

In a good design, the mechanical (rotational) inertia is made small and the resistance to rotation (damping) negligible. This fact, associated with close coupling in most fluids, enables turbine meters to follow rapidly changing flows. Responses are of millisecond order so that transients and low frequency pulsating flows can be followed faithfully.

In air speed measurement, the rotating-cup anemometer (this term is usually reserved for air flow meters) works on the same principle. Shaft rotation provides signals via magnetic sensing. A recent version uses a magneto-resistor to sense the movement, as this has greater sensitivity. It has already been stressed that flow is difficult to measure accurately and tests of anemometers and more advanced instruments indicate that the rapidly changing nature of wind can lead to considerable errors in the former.

Drag-plate flow meters use a plate suspended in the stream. This is constrained by springing, for example by means of a torsion hinge, cantilever or spring-loaded lever. Its movement is monitored by a displacement transducer. A simple unit made in the National Engineering Laboratory in Britain is shown in Fig. 5. If the spring system is stiff, movements will be small but response high. The unit shown moves only 100 μm and can follow pulsating flows to 100 Hz.

There are other ways to make use of the momentum of the fluid. One intriguing device uses it to alter the precession torque of a gyroscope. The liquid is piped through a loop made in a plane perpendicular to the axis of flow (that is, the fluid moves in a circle where the flywheel would normally be). The flow through the loop acts as the flywheel, producing angular momentum in the same way as a flywheel. If the loop is rotated, the gyro torque output is a measure of flow rate so the developed instrument nods the loop about a mean position to develop the torque as an ac signal which is then converted to dc form.

Another device uses two turbine blades of differing pitch that are joined together on common bearings by a torque spring coupling. They rotate in unison in the flow but as each experiences a different torque because of the blade pitches, they take up a relative angular position different to that at zero flow. Sensors at both blade positions deliver two trains of pulses. Their phase difference is a measure of mass flow.

TIMING SYSTEMS

When only a single measurement of flow rate is needed, the simplest way is to drop an identifiable marker in the stream. A ship's log is used at sea in this way to measure relative speed. Similarly, in channels or pipes a tracer can be injected. Common salt is often used in clean rivers. In polluted waters lithium salts may work. The gulp method dumps the tracer into the flow. Downstream samples are collected at known time intervals and analysed for salt concentration. Plotting values against time shows when a maximum is reached and this plot is related to flow rate. Suitable tracers are chemicals, dyes or isotopes, the latter having limited use due to health hazards.

When continuous measurement is needed, such tracers are usually impracticable, but the concept can still be used with heat or motion bursts that are generated in the fluid. Although the principle is simple, the signal levels are small and they exist in the presence of severe noise signals of the same form. The use of cross-correlation techniques has been proven capable of measuring the time delay for heat perturbations to pass between two points despite poor signal/noise ratios. As an example of this very recent method, a thermal flow meter is now explained.

Fig. 6 shows a schematic of a cross-correlator thermal flow-meter showing the major fluid and electronic components. At a convenient point in the flow is a small heater which is energized by a signal generator providing a pseudo-random noise signal. This imparts to the fluid relatively small quantities of heat in a time sequence resembling random heating. (In practice this generator is binary in nature providing only two levels of signal to the heater but in a continuing random time sequence – it is termed a pseudo-random binary sequence, P.R.B.S. generator). Down stream are two, fast-response, temperature sensors a metre or so apart, so that each receives the same thermal fluctuations but at different times. Merely examining the temperature sensor signals would reveal little more than a noise signal with no clear definition of the original input to the heater. Therefore a process of correlation is used to recover the buried signals.

Correlation can be visualized by considering two identical complicated optical patterns formed on film transparencies. When the two are exactly overlaid, the maximum amount of light transmission occurs. If we now misalign the patterns, the transmission attenuates rapidly as the degree of misalignment is increased. In the correlation of electrical signals the signals are time variables rather than space variables. In essence, the two signals are multiplied together and the multiplicand signal then averaged. This is repeated many times with different time delays. At a processing time-delay equal to that of the time taken for the fluctuations to travel from one probe to the next, the correlation output will peak quite sharply. In Fig. 6, m(t) and n(t) are the two temperature signals. The multiplier unit gives $m(t-\tau).n(t)$ which is integrated to produce the output signal. The time-delay unit provides delay increments. In continuous signal monitoring, the delay unit is tracked to keep the cross-correlation output maximized.

Commercial correlators are available, but being general purpose instruments, they are usually expensive. Less expensive units (such as that shown in Fig. 7) can be made to suit specific cases such as this example.

Any signal that can be made to perturb the existing state of the fluid and then be detected may be used, the essential factor being that the pseudo-random signal marking the flow must retain its spatial form between the markers. In slurries and gaseous suspensions there is often no need to add a signal since the medium has inbuilt patterns due to voids or denser particles. In these cases, it is only necessary to sense the effect at two places using capacitive sensors for example. (Surface velocity is similar to flow measurement, surface irregularities providing the signal. For example, in aerial photography the aircraft ground speed is needed. One solution is to cross-correlate the appearance of the ground seen by two sensors viewing at different places along the flight path). A decade ago correlation methods were novel and in the experimental stages. Today they are used extensively as a routine procedure.

Fig. 7. Multi-purpose correlator unit combined with spectrum analyser. (Hewlett-Packard)

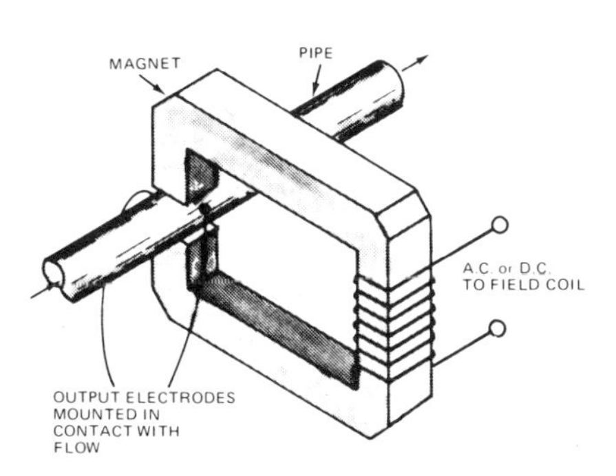

Fig. 8. Principle of the electromagnetic flow meter (an iron circuit is not always used).

ELECTROMAGNETIC FLOW METERS

To understand the operation of electromagnetic, E.M. for short, flow meters, it is necessary to look at the findings of the 19th century scientist – Faraday. The Faraday principle states that an electrical conductor cutting a magnetic field experiences a force acting upon it that is proportional to the rate at which lines are cut. In E.M. flow transducers, use is made of the voltage generated when a field is cut by a current carrying conductor. Referring to Fig. 8, the coil magnetizes the pole piece and the liquid (the conductor) moves through the field produced in the air gap of the iron. Some E.M. flow meters use two coils laid on the pipe walls. Electrodes are placed at the positions shown for that is where the generated voltage appears. Regardless of the state of the flow, the method indicates accurately for it relies on the bulk electrical and magnetic properties of the fluid in a given volume.

When electrolysis problems might occur, the field is excited with ac, either from mains excitation using a transformer, or from an inbuilt signal source as shown in the schematic of a flow meter in Fig. 8.

E.M. meters have found use in the measurement of the flow of liquid sodium, saline solutions, seawater, blood, mercury, electrolytes such as plating solutions and, of course, water. A general guide is that the conductivity of the fluid must be greater than 10^{-7} siemens/cm^3 for satisfactory signal generation. Tap water has a conductivity of about 2×10^{-4} siemens/cm^3.

The design of the coil and electrode shape is important. The field produced should be at least three pipe diameters in length so that the electrical shunting conductivity of the fluid just outside the field boundaries is made insignificant. The pipe of the meter must be non-magnetic – and have insulated readout electrodes (plastic is an obvious choice). If possible, the pipes either side of the meter should be non-magnetic and insulating for a short distance each way. In precision installations, the meter should be calibrated. Signal levels are small (order of millivolts) but the use of ac achieves good signal/noise ratios as the thermoelectric and electro-chemical potentials are not amplified in the detection equipment.

E.M. flow meters have been commercially available since 1950. Ranges available cover 3mm to 2000 mm diameter pipes. The smooth bore design makes them suitable for slurries, pastes and liquids having solids in suspension.

ENERGY BALANCE DEVICES

Hot wire or hot film anemometers consist of a fine resistance element suspended so as to make good contact with the gas flow. The element is heated and its resistance measured in a Wheatstone bridge circuit. Convective losses from the element, due to the gas flowing over it, depend on the velocity of the flow so the temperature of the element will stabilize at a value where the energy lost to the flow equals that supplied to the sensor.

Three basic forms are in use – constant current through the element, constant temperature of the element or, less common, constant resistance ratio between two elements. In the first, a constant current is fed to the elements (see Fig. 9a), so the element temperature drops as the flow increases. Flow is measured indirectly as the resistance of the element which depends upon its temperature. The second method alters the current so as to maintain the element at a constant temperature (as shown schematically in Fig. 9b), – the current is a measure of the flow rate.

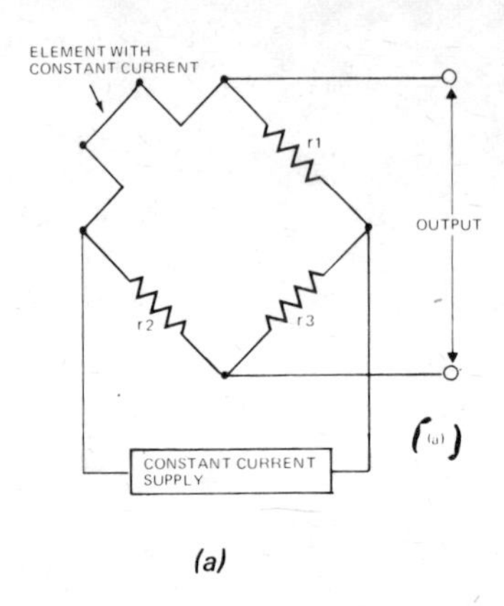

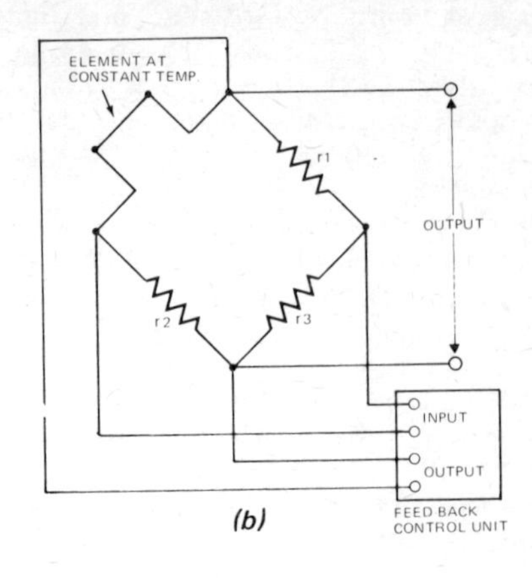

Fig. 9(a). Constant current hot-wire anemometer bridge. (b) Constant temperature operation requires feedback to hold the element temperature constant.

If the flow does not have a steady temperature, but fluctuates, the above methods will not be accurate since energy loss depends on gas temperature. The constant resistance-ratio method is a way to reduce temperature errors. It has two elements, each fed with a different current. Their voltage ratio is maintained constant with a feedback system where voltage is taken as a measure of the flow rate.

The elements must be small to obtain a fast response; an example is tungsten wire, 40 μm in diameter. Deposited films are also used. In turbulence studies, the system response needs to be as high as 50 kHz. Hot-wire anemometers have been developed that easily cover the need – having response times of 10 μs. Obviously, such delicate probes can only be used where the flow is clear of particles and contaminants that may alter the convective film coefficient and it is necessary to calibrate these devices in-situ.

The boundary-layer flow meter also makes use of heat transfer and energy balance but in a very robust way – no delicate probes are used but the gain is at the expense of response. A heating coil is wound around a thin heating pipe as shown in Fig. 10. This produces a temperature profile across the cross-section of fluid in the pipe. Fluid near the wall is the hottest because this, the boundary layer, is not moving. The temperature drop across the layer is related to its thermal conductance which, in turn, depends upon the (mass flow rate)$^{0.8}$. One probe, therefore, measures the wall temperature, the other the fluid temperature on the other side of the layer. It is not necessary to place a probe in the stream for the centre temperature can be measured up-stream before the heater. In use, the heater is adjusted to keep the temperature drop constant. In this mode, heater power level is a reasonably linear measure of mass flow rate.

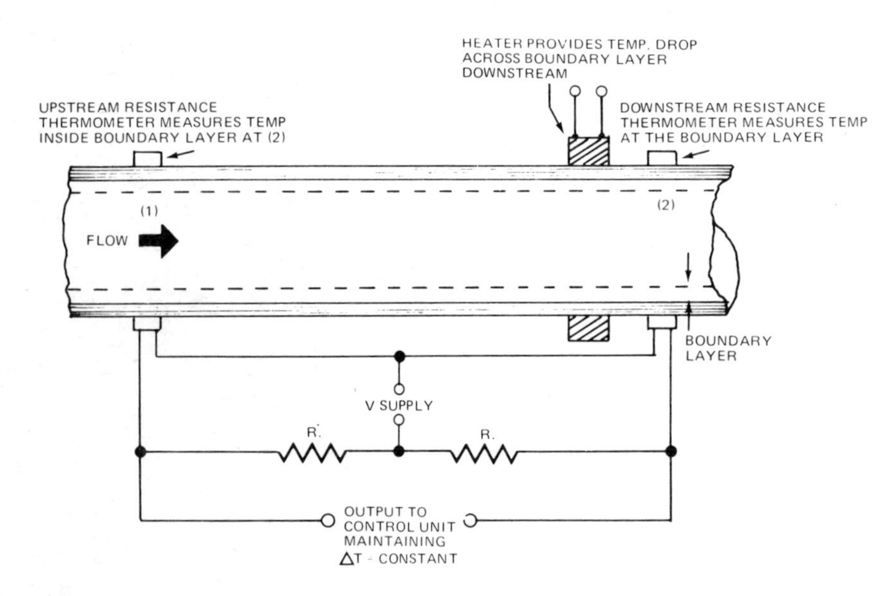

Fig. 10. Cross-section through a boundary-layer flow meter.

DOPPLER FLOW METERS

Energy radiated through a fluid in motion will reach a given point elsewhere at a later time depending upon the rate of energy propagation and the velocity of the fluid in the same direction. Another significant phenomenon is that the frequency of the received signal will be altered from that of the sending source due to the Doppler effect. This can provide greater resolution than straight transit-time measurement. Ultrasonic, radar and laser radiation are employed but the hardware differs in each system due to the different wavelengths.

ULTRASONICS

Piezo-electric crystals, one for transmission and one for reception, are positioned in the wall of the flow-meter aligned in direction of fluid flow. These provide radiation and detection of the acoustic pressure waves that are launched into the fluid (as shown in Fig. 11a). When the fluid is liquid, there is no problem in obtaining an efficient energy coupling, but for gases, it is a more formidable problem.

In the simplest arrangement (Fig, 11a), the upstream crystal transmits a burst of high frequency (at typically 10 MHz) that is received down-stream. The transit-time will vary in relation to fluid velocity but this measurement also depends upon the (acoustic velocity)^2 so variations in the speed of sound greatly affect the accuracy. Furthermore, the delay time can be very small, making it hard to obtain resolution. Usually a more complex arrangement is used which has two systems acting in opposite senses, as shown in Fig. 11b. Each loop resonates because the received signal is used to send the next pulse burst (called the sing-around technique). When the fluid is stationary, the frequencies will be the same but with flow movement one frequency goes down, the other up. The beat frequency formed by comparing one with the other is a direct measure of flow velocity.

More recently developed methods make use of the frequency shift due to the Doppler effect. Energy received elsewhere (or sent back) to the source will be of different frequency to the source, so frequency comparison similar to the above gives flow rate. A blood velocity ultrasonic Doppler meter is shown in Fig. 11c; note the gel used acoustically to couple the crystals to the artery wall.

Before leaving ultrasonics, it is worth mentioning that ultrasonics can be used to detect leaks in pipes. Gas passing very small orifices tends to resonate in the range 30-40 kHz. Ultrasonic detectors will detect this, often at a considerable distance.

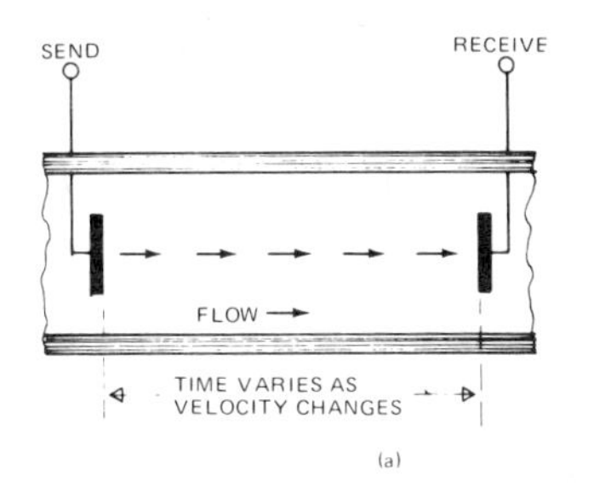

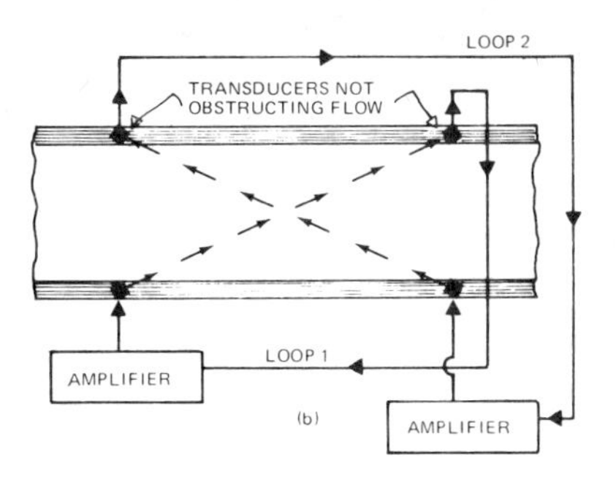

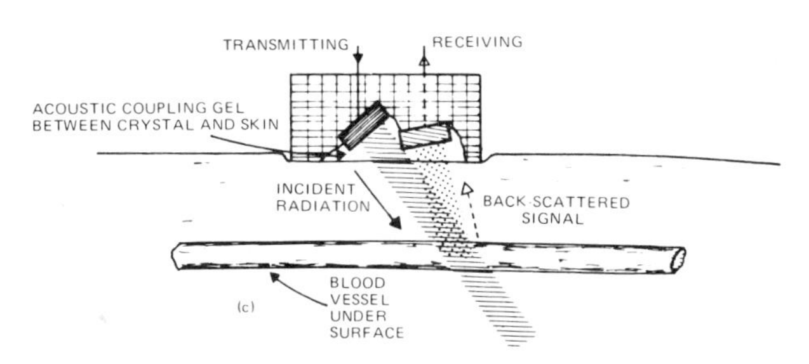

Fig. 11(a). Simple ultrasonic transit-time variation method of determing flow rate. (b) Double resonant-loop method reduces errors. (c) Ultrasonic Doppler used to monitor blood flow.

RADAR

The Doppler principle has also been applied with radar sources in order to provide a frequency variable signal related to flow. The availability of miniature self-contained C.W. radar sets (such as the Royal Radar Establishment unit which is camera size and runs from batteries) has made radar Doppler a reality in industry. The unit is arranged to look at the flow at a slight angle (sand, water, films, granules, liquids with air bubbles, water drop sprays and surfaces of sheet material have been measured). The same antenna acts as a radiator and receiver detecting back-scattered energy which is compared with the source frequency. The signal does not have constant amplitude nor does it exist continuously, hence reasonably sophisticated electronic equipment is needed to determine the frequency difference. Fig. 12 shows a schematic of the instrument in use — along with typical waveforms from various materials as published by the Institute of Measurement and Control. Filtering is used to improve the signal/noise ratio but it has been found that the best results are achieved by using a plastic zone plate (a disk having annular rings turned in it) between the instrument and the surface, rather than using electronic filters after the detector.

LASER VELOCIMETERS

One of the applications of laser sources is for measuring velocity with great accuracy and large dynamic range. (In principle, it is possible to monitor a range of 10^7:1). Since the early 1960's, the technique has been improved and is now commercially available. The laser beam is split (see Fig. 13), to provide two coherent sources what interfere optically at the point where velocity is to be measured. (It is necessary to make an optical comparison since interference is not possible electronically, and we do not, as yet, have detectors for such high frequencies). A sensor viewing this point sees a small circular fringe pattern that varies in amplitude as

Fig. 12. Doppler radar unit compares reflected signal and source frequency to provide frequency variable signal related to flow rate. Waveforms from a number of materials are shown.

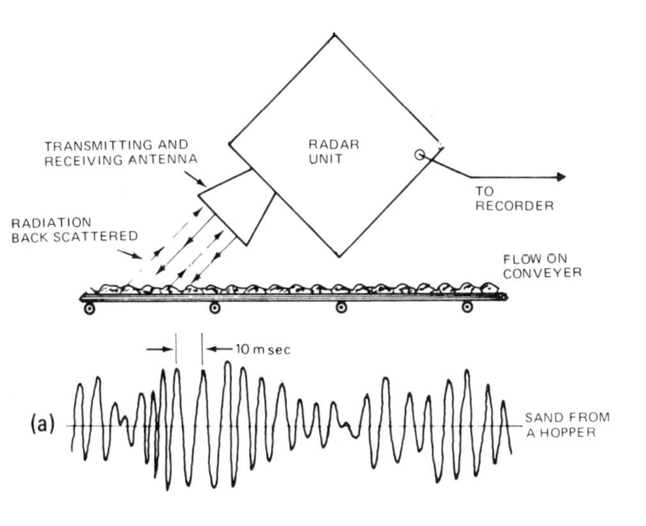

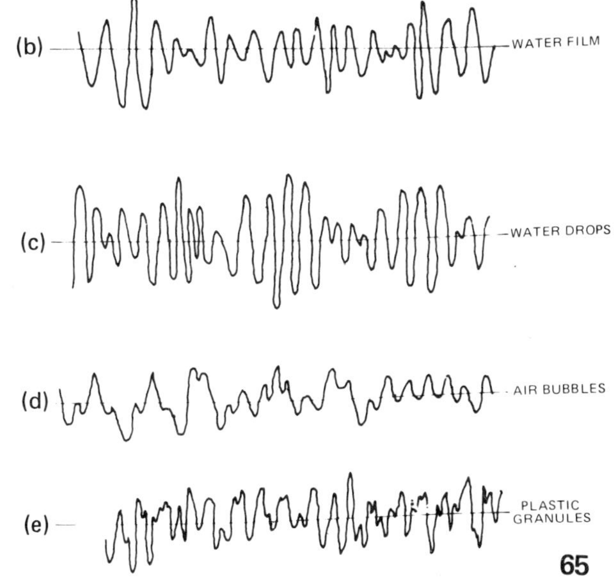

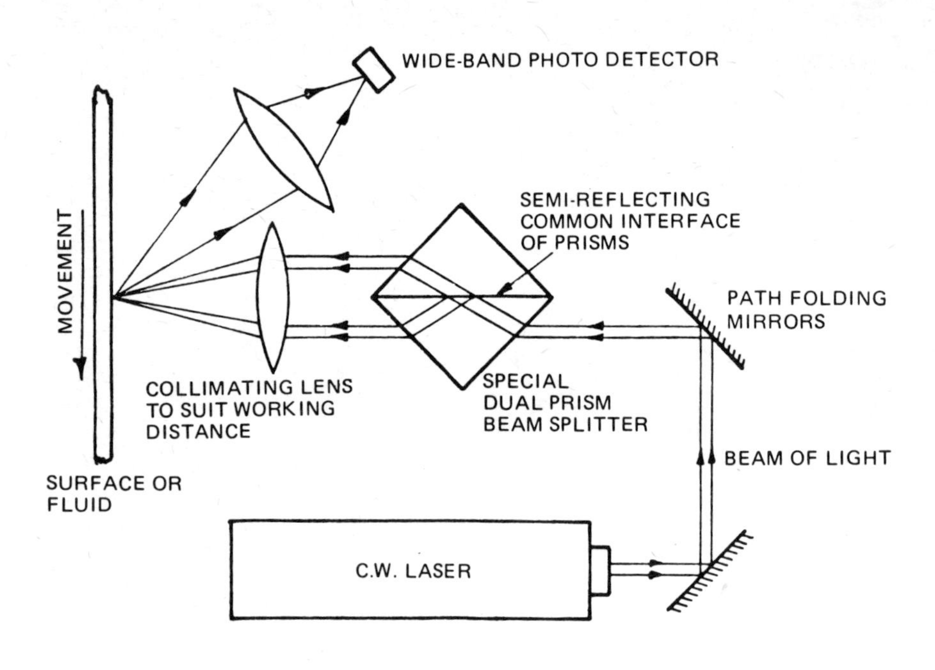

Fig. 13. Laser velocimeter. The laser beam is split to provide two coherent sources that interfere optically at point where velocity measurement is required.

scattering changes. If the medium is moving across the field of view, the sensor detects passing fringes and produces short bursts of signal. The period of the cycles in a burst is a measure of the velocity. Extensive electronic processing is needed to produce accurate flow measurements on such vague signals. The main advantage of laser flow meters is that the velocity of a volume of fluid only 10^{-3} mm^3 is viewed. The method is most useful in turbulence and profile studies. It is essential that some, but not many, scattering particles exist to provide a signal for the detector. Often air bubbles or a colloidal solid are injected to enhance the signal strength.

MISCELLANEOUS METHODS

In a cryogenic flow meter, the gas is made to flow through a thin flexible mesh held across the stream and parallel to an insulated electrode. As the mass flow velocity increases, the mesh is pushed closer to the electrode, changing the capacitance between the two. This is used to alter the frequency of an oscillator providing a frequency output form of signal.

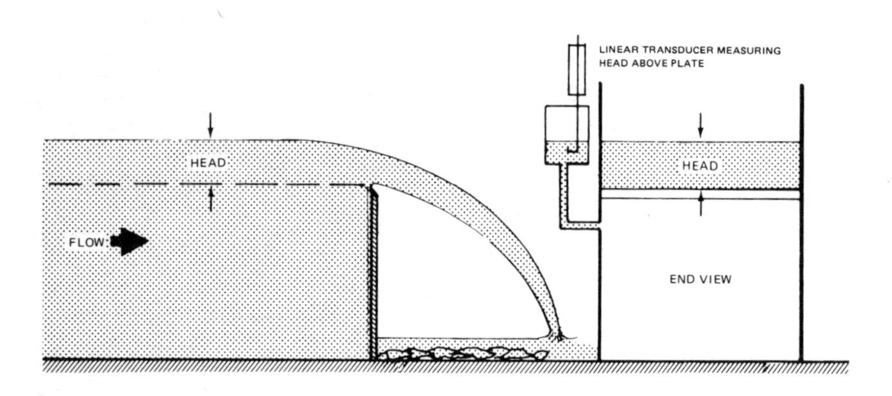

Fig. 14. This technique may be used to measure large volume flow in open channels.

When large volumes of liquid move in open channels, flow can be monitored by lowering a flow meter into the stream, but there is a simpler way in permanent installations. Theoretical considerations of the shape of flow across plates or weirs leads to equations that relate flow rate to the shape of obstruction, fluid constants and height of water above the plate. This is the same concept as orifice plates, etc. but the obstruction is a continuous circle. The relationship is not linear (quantity depends on $height^{3/2}$) and shape is vital. Standards are available so that channels built to specification (see Fig. 14), can be made that give a known calibration. A float is arranged to operate a linear or rotary displacement transducer. Plates and weirs can also provide differential pressures.

The oscillating-fluid flow meter is the last to be discussed. If fluid passes through a cavity of the correct design, it will oscillate at a frequency related to velocity (and temperature – remember the fluidic temperature sensor). Oscillation is induced into the stream entering the flowmeter by passing the fluid through vanes arranged to impart a swirl inside the cavity. Oscillation frequency is detected by monitoring the fluid temperature at a point in one commercial unit. This method has a fast response – 1000 Hz is claimed and is linear to about 1% of chosen full-scale range. It operates both with gases and liquids providing they are homogeneous and the cavity is designed to suit the fluid to be used.

FURTHER READING:

"Handbook of transducers for electronic measuring systems", H. N. Norton, Prentice-Hall, 1969.

"Symposium on the measurement of pulsating flows", April 1970, Institute of Measurement and Control, London.

Extensive bibliography (to May, 1972) of blood flow measurements available from Parks Electronics Laboratory, Beaverton, Oregon, U.S.A.

"Measurement systems: application and design", E. O. Doebelin, McGraw-Hill, 1966.

"Fluid mechanics", V. L. Streeter, McGraw-Hill, 1958. (Background text for basics of fluid flow).

"Symposium on measurement and process identification by correlation and special techniques" January 1973, Institute of Measurement and Control, London.

CHAPTER 10

FORCE, WEIGHT AND TORQUE

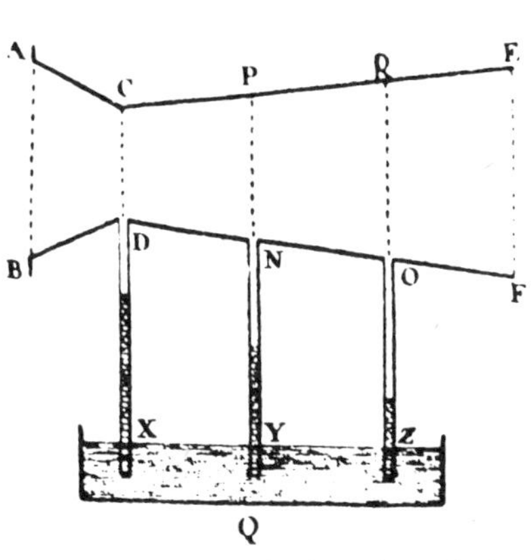

Manometer set-up as drawn by Venturi in 1797 for measuring pressure in a wind tunnel.

Prototype kilogram once used to define mass in the United States of America.

Physical objects have a property of mass as well as size. This is a fixed value for an object (provided time has no influence). It is the magnitude of the gravitational attractions upon it that decides its weight (weight = mass x gravity). An object exerts a force upon its support due to the gravitational attractions. When zero gravity exists there is no such force — articles float in outer space because the gravitational pull of the heavenly bodies is negligible. A force can also be exerted by an accelerating mass (as Newton realised with his force = mass x acceleration equation) in the absence of gravity. Gravity is but one form of acceleration. A force can also be created by devices such as springs or magnetic attractions which do not rely on acceleration or gravity.

Basically then, the weight of an object is decided by our knowledge of the gravitational pull acting on it. Fortunately, variations in gravity with time and place (on the Earth at least) are small compared with the precision needed in most commercial applications and a fixed value, or even a total disregard for it, suffices. (For example the roughly 12 hr period gravity variations due to the effect of the changing orbit of the Moon are roughly one part in 10^7.) The standard value of gravitational acceleration is 9.80665 m/s^2.

Units of mass, weight and force in existence are varied. The Imperial systems' pound and slug (an engineering term used for mass as opposed to weight) have been confusing to the uninitiated. The distinction between mass and weight is often not made.

Metrication dictates the use of the kilogram kg as the mass unit and the Newton N as the force unit. As force and weight are often synonymous it has been usual practice to qualify force by the use of the letter f as in lbf but now only Newtons N and kilonewtons kN are to be used for force with the gram g, kilogram kg and tonne t being reserved for mass.

Not always is it the total force of a system that is of interest. Instead it may be a derived quantity. Pressure is the force exerted by a distributed force on a unit area of surface. Torque is a variable acting about a pivot joint that is created by a force effectively acting at the end of a lever arm.

Common Imperial pressure units are pounds per square inch, psi, or inches of mercury Hg or water H_2O. (A column of liquid exerts an absolute pressure at the bottom depending on its density and the height of the column). In the new metric system the unit of pressure is the pascal Pa (Pa = N m^{-2}). Older units were kilograms per square centimetre or metre and centimetre height of water or mercury. In meteorological measurements of atmospheric pressure the millibar mb is commonly used — weather charts express pressures of highs and lows in hundreds of millibars mb. A standard atmosphere is now exactly defined as 101.325 kPa, but millibar units will still be used in weather records. A pound per square inch equals 6.89 kPa.

To confuse the issue further, vacuum pressures (those less than atmospheric in general) often use a unit 'torr' that is the pressure exerted by 1 mm of Hg at 0°C. In this range it is also acceptable to use mm Hg and μHg units. Specific mention is made of vacuum gauges for they measure low pressures by quite different methods to the above atmospheric pressure devices.

To conclude this brief resume of units there are three forms of pressure definition. Firstly it can be expressed in an absolute sense, pure vacuum being zero pressure. In absolute terms atmospheric pressure is the familiar 14.7 psia or the 101 kPa mentioned above. (Note the use of 'a' to denote absolute). Secondly, a gauge pressure can be used, psig for instance, where atmospheric pressure is used as a reference datum of zero pressure. Thirdly, pressure might be the difference between the unknown and a convenient datum pressure other than atmosphere. This is the differential pressure — psid is how it is denoted. If the pressure is stated in the column height form it could be designated as in.wg (water gauge) or inH_2O

measuring inches of water.

Torque is now correctly expressed as newton metres N m but the many older units such as pounds force-inches, ounce-inches, dyne-centimetres, kilogram-metres and others will be in use for a considerable time to come. They should not, however, be used in newly-prepared documents describing new products.

THE STANDARD OF MASS AND CALIBRATION OF FORCE

As force, weight, pressure and torque depend on mass it is appropriate to discuss the means by which the unit of mass is standardised.

Like all standards, that of mass is man-made — there is nothing about our everyday experience of nature that is regular enough to suggest its use as a standard. The aim, therefore, has been to provide commerce and science with a certain piece of substance that is the sole mass standard. The earliest mass standard seems to be the beqa used by the Egyptians around 3 800 BC. it was a cylindrical object with rounded ends (perhaps to ensure no corners could be knocked off?) that weighed around 0.2 kg. It is said our Troy weight system developed from this. As civilisations flourished independently in those times there were also other standards. Today the standard is universal and is a lump of platinum-iridium that is held by the International Bureau of Weights and Measures (BIPM) in Paris. This object defines the prototype kilogram — all others are substandards and are calibrated by comparison with this.

Many modern standards are now based on atomic phenomena (a verbal definition of an apparatus enables any group to construct its own standard to the same precision without need to actually inter-compare the two — atomic phenomena give excellent results) but to date it is not possible to relate the prototype kilo with the fundamental mass unit (1.66×10^{-27}kg) to a precision equal to the current arrangements of merely weighing the unknown and the standard. Even a comparatively simple beam-balance can be used to compare similar masses to within parts in 10^9 when used with adequate precautions. If the masses are of different material the task is not so simple. Firstly, the bouyancy of each in a fluid will be different due to their different volumes, and secondly they each will have a different quantity of absorbed gas. The use of vacuum weighing to avoid the bouyancy problem only introduces a problem by removing the absorbed gas that is regained when the vacuum is released. The practical solution (standards must be a practically useful arrangement) is to compare different masses in a

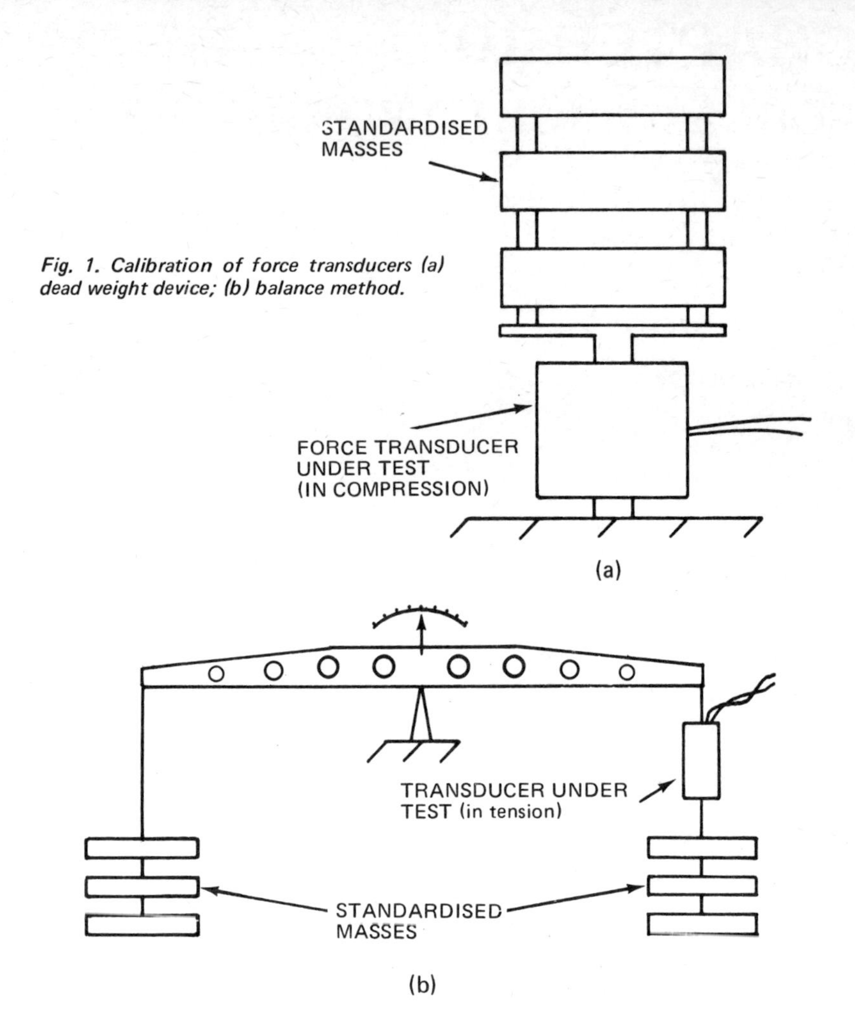

Fig. 1. Calibration of force transducers (a) dead weight device; (b) balance method.

standard air pressure.

In use, a standardised mass will exert a defined force due to gravity. Force transducers are therefore calibrated by putting them in series in a force loop. The two ways of doing this are shown in Fig. 1. Deadweight testing rigs go as high in capability as 100 000 000 N. Balances usually are used to calibrate to 10 000 N.

FORCE TRANSDUCERS

The need to determine the force acting in a mechanical loop arises in the calibration of pressure transducers, in the testing of civil structures where stress levels must be known, for weighing both static and moving loads, in the testing of automatic machine-tool structures, as the basis of accurate electric current determination and in force-balance devices mentioned earlier.

Unlike other variables there are comparatively few forms of force measuring transducer. The main principle employed makes use of an elastic mechanical member that deforms proportionately with loading. A secondary output transducer monitors the resultant displacement.

The design-aims for the elastic member are to achieve a linear movement of adequate magnitude that suits the available microdisplacement sensors. It must also have low mechanical hysteresis, a long fatigue life and be minimally influenced by environmental factors such as temperature. A few of the force sensing elements in common use are shown in Fig. 2. In each arrangement the sensor is usually placed at the position of maximum stress. Resistive strain gauges, inductive and capacitive sensors are generally used. The complete device is termed a load-cell. In some designs a solid-block is machined out to produce areas of considerably high stress to enhance the sensitivity. In the measurement of dynamic force systems a compromise must be made between the load cell spring rate (a more elastic cell gives a larger strain signal) and the sensitivity obtained (a less elastic cell results in higher resonant frequencies but gives a smaller output signal).

Piezo-electric ceramics provide an electrical output directly from the application of force across the device. During the initial period of force variation, charge will flow to balance the energy but as there is only a fixed amount of energy available for a given force x distance product the

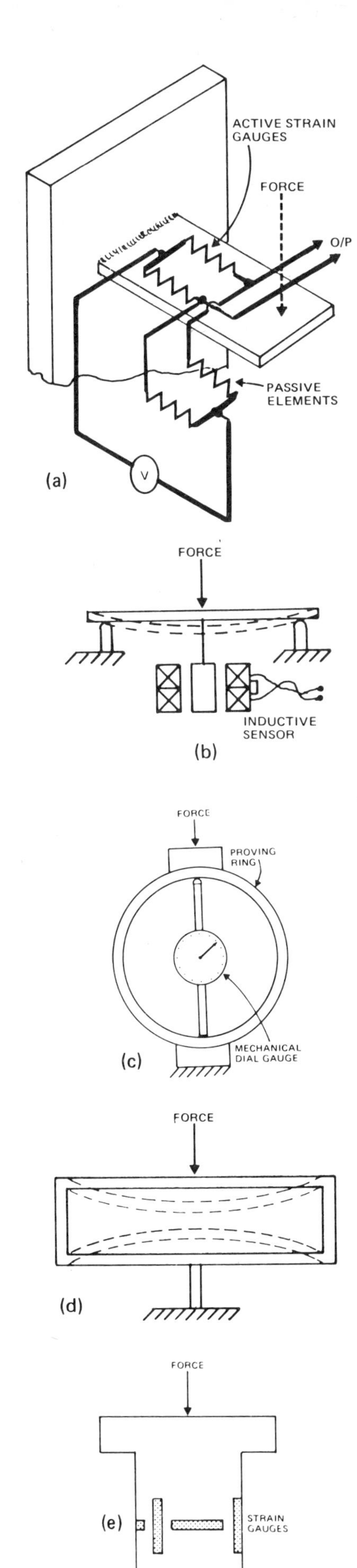

Fig. 2. Elastic elements used in force transducers. (a) cantilever; (b) simply supported beam; (c) proving ring; (d) disk cell; (e) solid or tube column.

signal eventually drops to zero. To extend the duration of the signals (to give the piezo device quasi-static performance) an extremely low-leakage amplifier is used to read the charge flow. When measuring rapidly varying alternating forces, as for instance in machine-tool vibration analysis, there is a continual energy supply and the high input impedance amplifier is not as vital. The force transducers are extremely stiff so little deflection occurs across them.

Many instruments rely on springs to provide a calibrated balancing force that can be used to measure an unknown force. There is a basic difference between such dynametric devices and force assessment by the use of weights.

The weight of an object determined by the absolute method using a beam-balance will be the same regardless of the value of gravity for both masses are equally attracted. A spring devised determination, however, will vary with gravitational changes, for the spring provides a constant force–distance relationship. (This is the basis of many sensitive gravity-meters).

CHEMICAL BALANCES

Beam balances have been in use since the earliest times and even a crude arrangement can be used to compare weights to high precision. In the 18th century the rise of research interest in chemistry created a continual need for better balances. Joseph Black reported results of a chemical weighing to the nearest grain (0.065 g). By the 19th century Handolt was weighing 500 g loads to several micrograms.

The simple beam balance has been continually developed to reduce the errors in knowledge of the length of the arms, to increase the sensitivity by reducing frictional errors and to simplify the balance adjustment and reading arrangement. Modern balances often make allowance for bouyancy effects and are quite different in appearance from the traditional beam balance. Counterbalancing loads are sometimes applied by front mounted knobs and this provides for simple use and direct readout of value.

Microbalances measure small loads to micrograms resolution, (they are, however, not often small in size). Most make use of a fine torsion-fibre or ribbon suspension that is twisted by the load. To determine the exact load, a counter-torque is applied to rezero the beam. Figure 3 is schematic of a commercial electro-balance. The sample is placed in one pan and a roughly equal weight in the other. To achieve balance the current in the coil is varied until the photocell displacement detector receives a standard illumination level. The process is easily automated to follow changing weights. A recently developed microbalance makes use of a magnetic field suspension system to support the load without mechanical restraint. The field-coil current needed

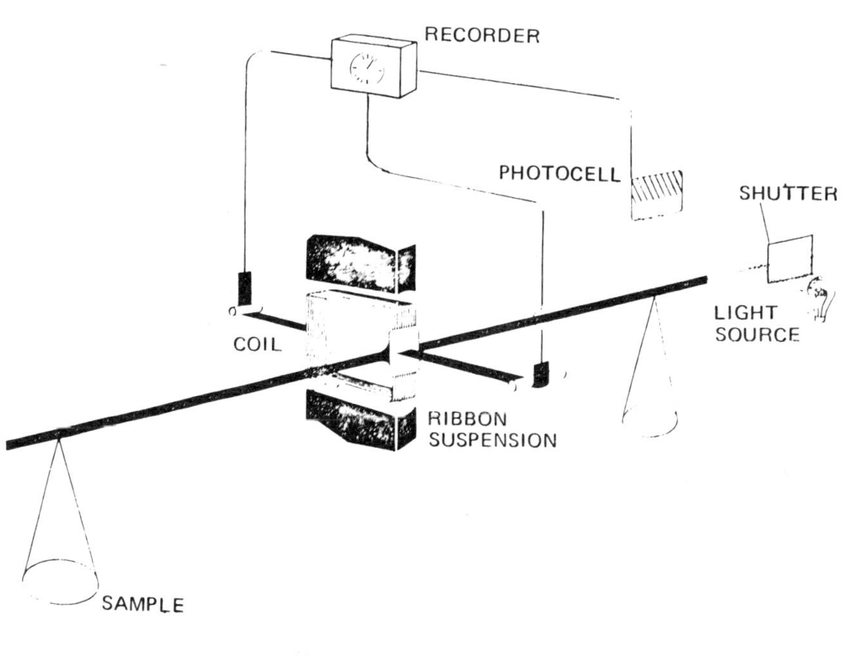

Fig. 3. Schematic of the Cahn Electro-balance.

to establish position balance is a measure of load. Modern microbalances can resolve weights of less than a microgram.

DYNAMIC WEIGHING

An automatic readout weighing machine consists of a platform operating on effectively frictionless pivots that bears down upon a load-cell. Details of a commercial unit are shown in Fig. 4. If the load is stationary, the reading is reliable, but a moving load, such as an animal or a travelling string of goods wagons on a railway or material on a conveyor belt, will produce a more complex signal such as that shown in Fig. 5. Averaging the value eliminates the fluctuations about the true mean but to obtain a fast response more sophisticated adaptable filtering is used. The power of such methods is shown by stating that a simple R-C low pass filter gives 200 times greater error than a weighted averaging method (patented by Avery) when used in an equal, short-time period.

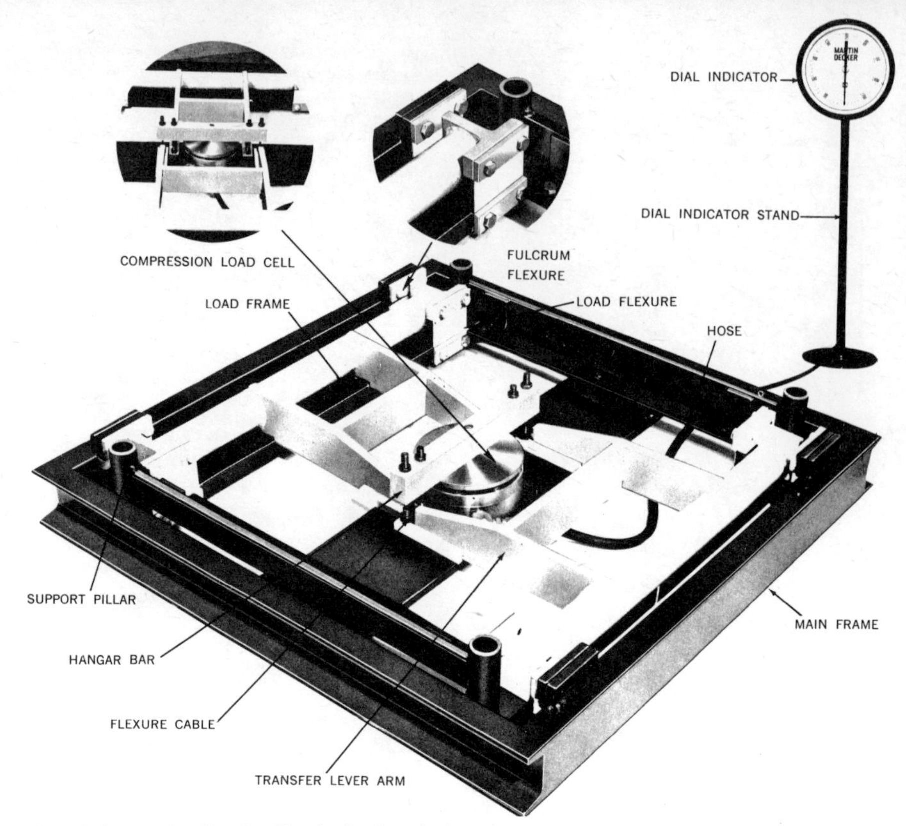

Fig. 4. Inside details of a Martin Decker deck scale.

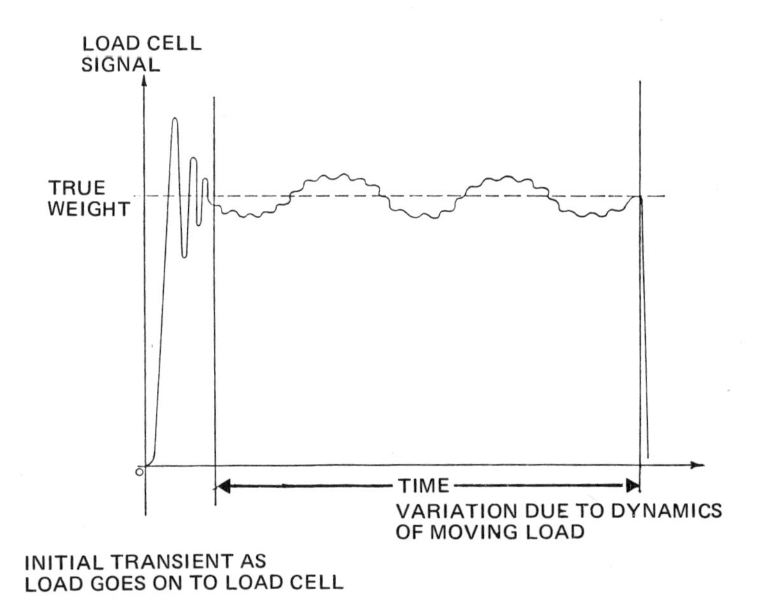

Fig. 5. Typical weighing signal produced by rolling stock passing over a battery of load cells.

PRESSURE SENSORS

The range of pressures that exist is vast and ranges from tens of megabars (10^8 atmospheres) down to inter-stellar space pressure (10^{-19} atmospheres). Consequently many techniques are used to suit the range.

As with force transducers the commonest procedure uses a mechanical elastic element coupled to a microdisplacement transducer. The main difference is that pressure transducers usually do not need the stiffness essential in force sensors.

Commonly used sensing elements are shown in Fig. 6. The limiting factor is usually the magnitude of mechanical hysteresis and perhaps bearing friction in the simpler devices.

The most general need for accurate pressure measurement around an atmosphere is for weather mapping and for elevation determination (barometric levelling). The aneroid barometer uses a stack of bellows to drive a pen or transducer. The best resolution instruments are termed microbarometers. The schematic of an advanced instrument is given in Fig. 7. Note the quartz helical Bourdon tube that is used because of the excellence of its stability and spring-rate. Microbarometers have been used to detect the pressure surges caused by nuclear test explosions on the opposite side of the globe.

Pressure sensors using an elastic element are not absolute and must be calibrated against those using a head of liquid as a reference, or by the use of dead weights driving a piston filled with liquid.

The standard mercury barometers, of the Kew or Fortin pattern, are devices in which the height of a mercury column is measured against a mounted scale. If the column is coupled to the pressure within a closed system via a pipe it is termed a manometer. These are used in abundance in hydraulic research as readouts for pressure difference flow meters and in aerofoil section investigation. Chemical reactions often must be monitored under precisely known pressure conditions. Manometers used in this work are often made more precise by electrically sensing the position of the top surface of the column. Auto tracking inductive sensors, such as that given in Fig. 8, and capacitive or optical sensing have been employed to resolve to micrometres. One micromanometer measures the surface height with an optical interferometer that looks at the mercury surface from above.

An analogue output signal may be undesirable from a pressure sensor. A sensor developed at the Royal Aircraft Establishment uses the force-balance technique in an oscillatory mode. From the schematic of Fig. 9 it can be seen that the driving coil forces the diaphragm against the pressure, opening the contact. This in turn de-energises the electrodrive, setting the system into vibration. The average duration of the contact dwell is the measure of pressure. This form of transducer can provide a time-duration modulated signal that is easy to transmit due to its binary nature.

Another transducer derives a frequency modulated output by monitoring the natural resonance of the sensing diaphragm. An electromagnet excites the diaphragm. A capacitive sensor monitoring the displacement is coupled to the excitation driver and the two oscillate

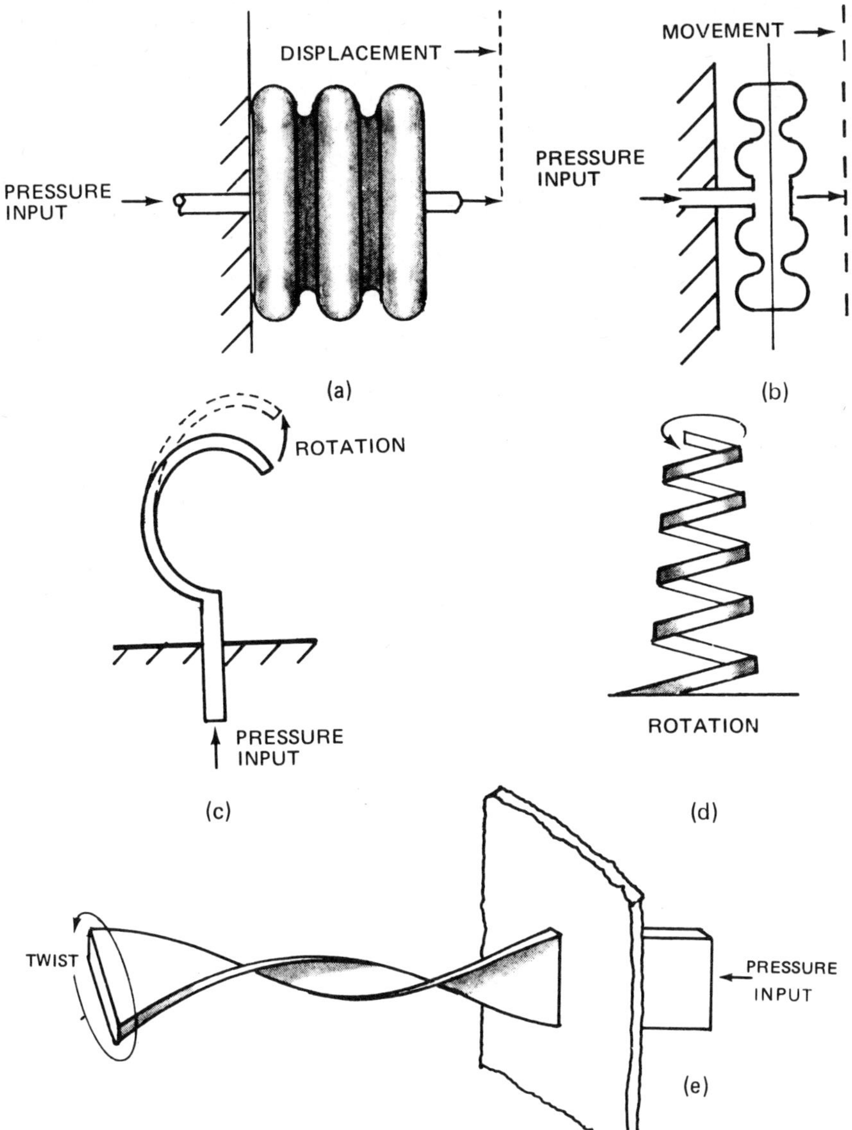

Fig. 6. Elastic elements employed in pressure transducers. (a) bellows; (b) capsule; (c) plane Bourdon tube; (d) helical Bourdon tube; (e) twisted Bourdon tube.

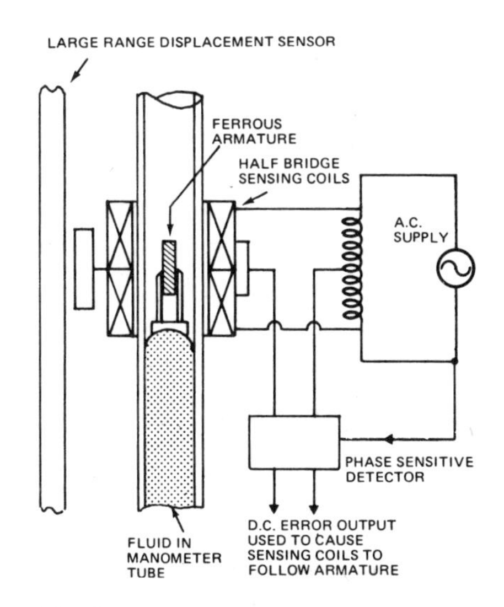

Fig. 8. The height of the liquid column is measured by slaving the detecting coils to follow the armature in this servo-manometer.

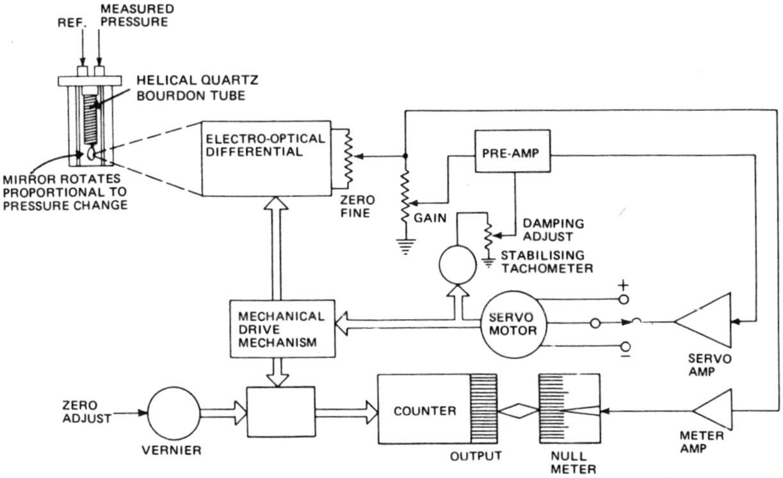

Fig. 7. Block diagram of the Mensor helical quartz pressure sensor having digital readout to 0.0005% of full scale.

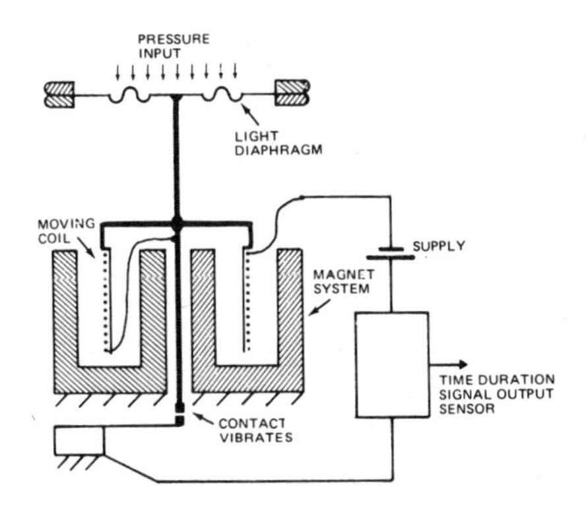

Fig. 9. Schematic of the RAE vibrating-contact, pressure transducer.

at the resonant frequency. As pressure changes so does the frequency.

VACUUM PRESSURES

From 1 mm Hg (1 torr) to an atmosphere is low vacuum, from 10^{-3} mm to 1 mm is medium vacuum. High vacuum extends from here down to 10^{-6} mm, very high from 10^{-9} to 10^{-6} mm. Ultra-high vacuum is a pressure less than 10^{-9} mm of Hg. Mechanical pressure gauges are often capable of measuring to 10^{-3} mm Hg but for better than low vacuum it has been necessary to employ quite different methods.

For the pressures of 10^{-4} mm Hg upward, the simplest vacuum gauge is the thermo-conductivity unit. The heat transfer from a heated sensor depends upon the mass of gas thermally coupling it to the container heat sink. The lower the vacuum the less the transfer. Earlier designs used a purely resistive sensor connected to a Wheatstone bridge to monitor its temperature via its resistance change, see Fig. 10, and this is commonly called the Pirani gauge. Another variation uses a thermocouple to measure the temperature. The most recently introduced models make use of a thermistor sensor instead. In this range the McLeod gauge – a type of manometer, provides an absolute method.

Below 10^{-3} torr, ionization gauges are used that rely on the ionization of the gas molecules to determine pressure. In the hot filament kind a heated cathode provides electrons that are accelerated by a grid to make impact on an anode, providing ions in the process. (The design is similar to a triode valve.) Positive ions are collected by an electrode and the ratio of ion to electron current is a measure of gas pressure. The Bayard-Alpert configuration, shown in Fig. 11a, has a small collector to reduce errors from unwanted soft X-ray generation. The anode collector can be replaced by an electron multiplier and versions using

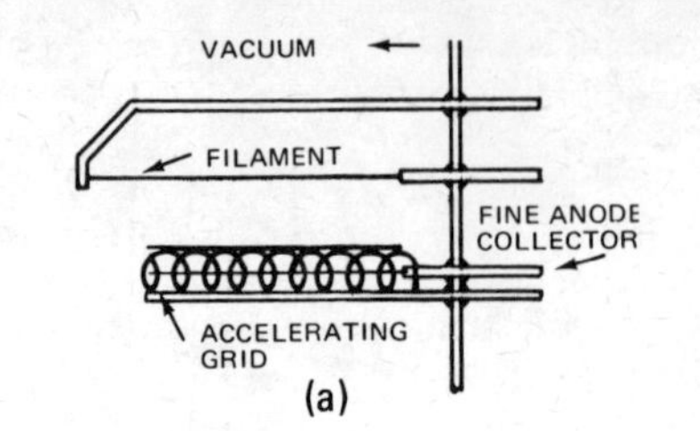

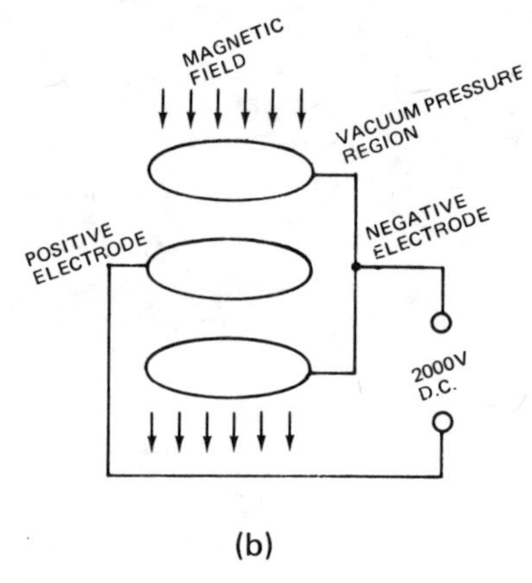

Fig. 11. Ionization gauges used to determine vacuum pressures. (a) This Bayard-Alpert sensor uses a hot filament; (b) In the Penning gauge a cold cathode is used.

this can detect 10^{-18} torr pressure changes.

The Penning, or Philips gauge, shown in Fig. 11b, is a cold-cathode, magnetic field, ionization pressure sensor. A discharge current is established by applying 2000 V dc between the anode and cathodes. The purpose of the magnetic field is to lengthen considerably the electron path (it makes them travel in a helical locus) enhancing the chance to form ions. The current drawn is the sum of electron and ion currents and is, therefore, not a linear measure of pressure. These are useful in the range 10^{-7} to 10^{-3} torr. The cold cathode is often preferred, for an unintentional vacuum loss will not destroy the cathode as would the heated version.

For pressures below 10^{-7} torr there are other methods. The Redhead gauge is also a cold cathode, magnetic instrument but the internal layout is different being based upon the Magnetron. It has the advantage of giving a linear ion-current to pressure relationship down to at least 10^{-10} torr.

The Knudsen radiometer gauge is occasionally used and is the only absolute method for the range 10^{-8} to 10^{-2} torr. A small mirror is suspended in the vacuum by a torsional filament that also supports a moving paddle-vane. Fixed close to the moving vane ends are two others that are heated. The momentum of the gas molecules passing near the heated plates causes them to bombard the unheated vane deflecting it in proportion to the gas mass present. The mirror forms part of an optical lever readout.

SOUND MEASUREMENTS

Acoustic propagation involves pressure wave travel. The above mentioned pressure gauges are of little value for sound pressure measurement due to the high frequency low pressure characteristics of sound waves. The detector, therefore, (the microphone) is designed to have a bandpass response including the frequencies of interest. Generally microphones have to couple an air medium so a comparatively compliant diaphragm is

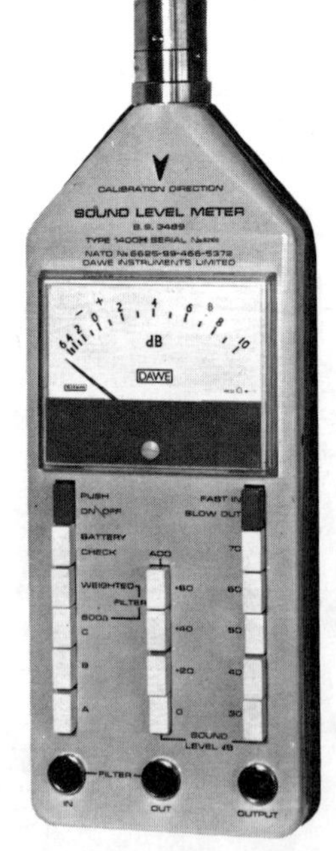

Fig. 12. The Dawe Sound level meter.

used to obtain an efficient acoustic match. Motion is transduced with a displacement sensor. Microphones employ piezo-electric crystal, capacitance and moving-coil principles.

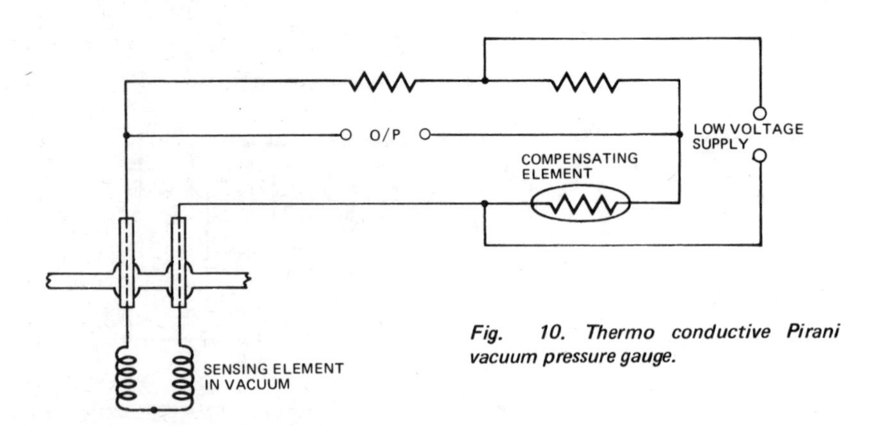

Fig. 10. Thermo conductive Pirani vacuum pressure gauge.

When making sound-level measurements it is necessary to use specially designed meters, Fig. 12. In these a stable characteristic microphone produces a basic pressure signal. This is processed to allow for the random-like nature of sounds and for the frequency characteristics of the human ear.

By definition a sound level pressure (SPL) of 0 dB is 0.0002 μbar this being the lowest discernible level for the ear (the pressure is 10^{-9} psi!). The compensated noise-level (that makes allowance for the frequency response of the ear) is expressed as the effective perceived noise level EPNL. The threshold of pain occurs at 144 dB. At 4 km, Concorde 002 produced 120 dB EPNL. The loudest noise produced is probably that generated by the NASA 16 m steel and concrete horn — 210 dB, which is 400 000 acoustic watts!

TRANSDUCING TORQUE

Torque is the product of a force acting at a distance, so for static shaft measurements a torque transducer is little more than a load cell driven via an arm (as shown in Fig. 13a). As the shaft is constrained it is not of any value for measuring rotating shaft torques.

A simple way to monitor the rotating shaft is to fix a strain rosette at 45 degrees to the axis (for this is the maximum stress direction) and connect the gauge to the bridge via slip rings as shown in Fig. 13b. Mercury-wetted rings have been used for extreme rotational rates; monitoring at 30 000 r.p.m. is quite feasible. A more sophisticated arrangement is shown in Fig. 14. Power to excite the bridge is induced by magnetic means, thus avoiding the problems associated with sliprings. A similar approach is to mount the power supply on the shaft along with the circuits and radio-telemeter the output. This has the advantage that the sending receiver and transducer needs no rigid position tolerance. It has been used in the measurement of tail-shaft torque in automobile research.

When the rotation is constrained, as with bearings at each end, a phase difference method can be employed. Detectable 'marks' (for inductive, capacitive or optical sensing) are mounted around the shaft at two points as shown in Fig. 15. Fixed proximity sensors generate two alternating waveforms which will have the same frequency but a varying phase-difference depending upon the amount of twist in the shaft. (This method is similar to a shaft encoder principle described in the earlier discussion on angle transducers). The major disadvantages are that there is no output when the shaft is stationary and that the frequency at which the phase comparison must be made varies with speed.

FURTHER READING

"Handbook of Transducers for Electronic Measuring Systems" H.N. Norton, Prentice Hall, 1969.

"Measurement Systems: Application and Design" E.O. Doebelin, McGraw Hill, 1966.

"Weighing Vehicles in Motion" A.C. Fergusen, Trans Inst. M.C. 1969, 2, 12, 214-222.

"Development of the Chemical Balance" J. T. Stock, Her Majesty's Stationery Office, 1969 (A Science Museum Survey booklet).

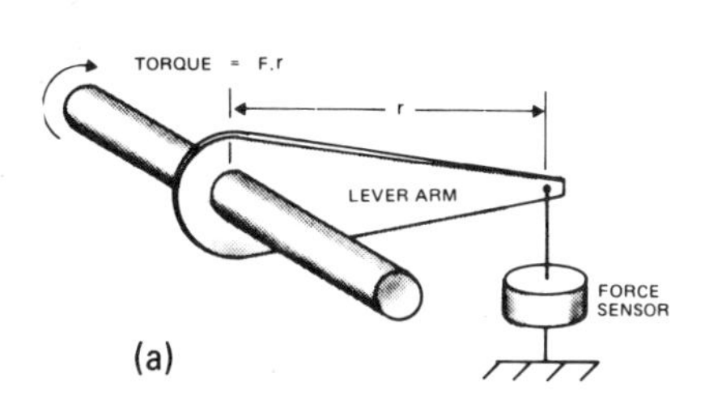

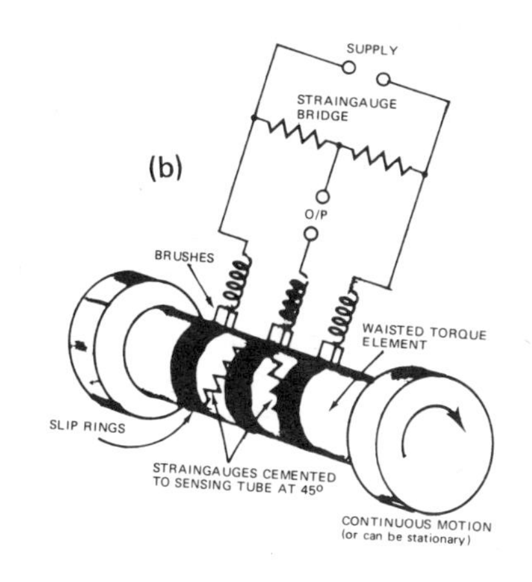

Fig. 13. Torque measurement. (a) Load cell and lever suffices if rotation is not needed; (b) Slip-rings enable the shaft to turn whilst monitoring torque.

Fig. 14. This Philips torque transducer supplies power and monitors the output of a strain gauge bridge via a non contact inductive coupling head.

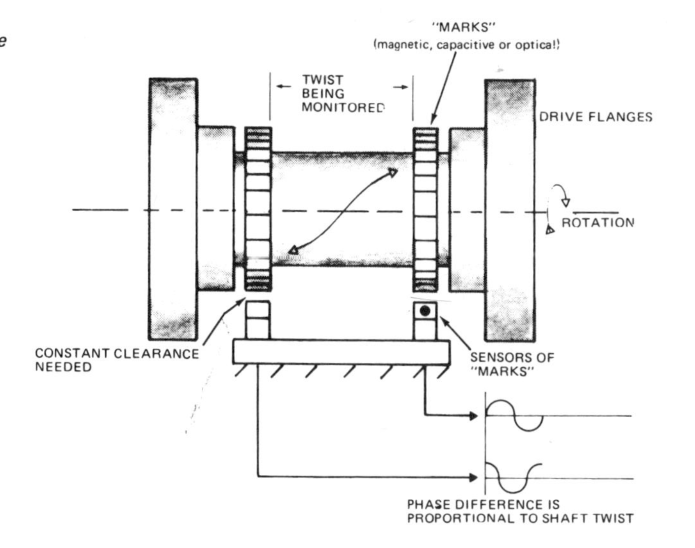

Fig. 15. Measuring the torque of a rotating shaft by the phase difference method.

CHAPTER 11
POLLUTION MONITORS — 1

Hardly a day passes without some mention of pollution, either in conversation or in the news media. Although much continues to be written on the harm being wrought on the Earth's ecological systems, its human inhabitants and to its resources, very little is ever said about the measurement problem itself.

Before pollution can be controlled, it must be detected, and that implies the need to measure. Some forms of pollution are obvious — litter that will not degenerate fast enough, thick smog, oil slicks, but many forms of pollution are insidious, going undetected until it is too late to take corrective action.

It is not the purpose here further to add to the literature on the problems resulting from pollution, but to provide a brief survey of some of the instruments used in the two main areas — the contamination of air and water.

CLASSES OF POLLUTION

Although we will not be covering all classes of pollution it is appropriate to mention them to put the discussion in perspective. One classification proposed uses five headings — air, water, land, noise and radioactivity. The first two are our main concern but as the last is also an important area of objective measurement it will be included.

Another way to regard the problems is by identifying the nuisance at source. Pollution is not a study of all contaminants but more of those of known annoyance to humans. In this way pollution divides into chemical impurities (lead, cadmium, oil, organochloride compounds, mercury, cyanide, sulphur and nitrogen compounds and hydrocarbons), biological waste and growth (sewage disposal and population excesses), radiation and ecological imbalance (wasteland deserts, loss of the natural insect and animal population control, disappearance of vital species) and noise (acoustic noise produced by transport, machines, industrial processes). There are others of a more subjective nature — litter, loss of clear view and the existence of unsightly buildings, but the task of measuring

these is most difficult for it is hard to define and qualify standards of allowable nuisance level.

AWARENESS OF POLLUTION

The settling and subsequent growth of cities began many thousands of years ago and this process naturally concentrated the elements of pollution. Fires, human waste and rubbish are concentrated geographically, and unless controls exist, the freely available air and fresh water soon become spoilt.

The Romans recorded their displeasure of the air of Ancient Rome. In 1273 the British instituted a not very successful smoke reduction programme: the penalties were harsh, however, for it seems a man was hanged for burning soft coal. London has been regarded the worst example of city filth for centuries. In Hogarth's time, 18th century London was much like his etching "Gin Lane" (shown in Fig. 1). In the 19th century Parliament often rose prematurely to escape the stench of the River Thames. But now London is one of the cleanest cities and has shown what can be done to eliminate pollution.

Plagues were common throughout Europe, annihilating as many as 65% of the European population in early times. There is little doubt that this was the result of throwing all refuse and sewage into the street. Tudor houses had the outward projecting upper storeys to assist this practice!

It was not until the 20th century that a real awareness of pollution appeared. In Australia a Smoke Abatement Act was introduced in 1902. An Alkali Act was introduced in Britain in 1906. But to have Acts and to use them are different things, and it was not until the 1950's that improvement became evident. The British legislated a Clean Air Act in 1956; Australia's was instituted in the 60's.

Motor vehicles have added to the problem enormously, providing air-borne carbon monoxide, solid

Fig. 1. "Gin Lane" – a famous etching by William Hogarth shows the highly polluted nature of life in London in the late 18th Century

hydrocarbons and lead in great quantities. In the U.K. in 1971 was consumed energy equivalent to 323 million tons of coal, a high proportion of this being liberated as CO, SO_2 and hydrocarbons. In the United States, (see Fig. 2), and Japan, the problem is even greater. The new, seemingly unrealistic, Congress Act to reduce vehicle emission is forcing design changes at the source of pollution. In this way the user pays the penalty — it is not passed on to others.

It has recently been estimated by the SIRA Institute that there will be an expansion of the market for pollution monitoring systems and devices from a current $600m to $3,000m in 1980. There certainly is room for improvement; for instance, few instruments exist that are within the price range of small companies and the domestic home. At present the accurate detection of most serious pollutants requires the use of a number of different, highly expensive instruments.

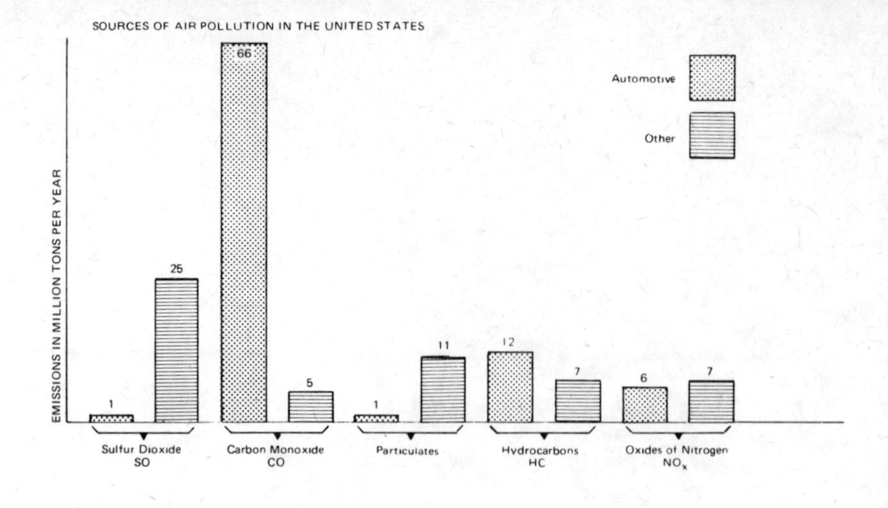

Fig. 2. A recent estimate of the quantity of eir contaminants in the U.S.A.

POLLUTION OF AIR

Let us now consider the contaminants of air and water. It will then be possible to study some of the transducers in use.

Air becomes contaminated mainly by man-made combustion processes. Fossil fuels (coal, oil and now natural gas) release gases and particles when the chemical process of burning takes place. The degree of harmful emission depends much upon the quality of the combustion process.

The main unwanted gases produced are carbon monoxide, carbon dioxide and sulphur dioxide. The first is physiologically dangerous for it can induce a deep fatal sleep without obvious signs. In lesser doses it produces severe drowsiness. It is, however, relatively easy to measure, especially at the exhaust of a vehicle.

Carbon dioxide, although not as harmful of life directly (as long as oxygen exists), does appear to have a far-reaching effect on the globe as a whole. This gas ends up in the upper atmosphere at an increasing concentration of some 0.7 parts per million ppm each year. Calculations indicate that a doubling up of the current concentration of around 300 ppm will reduce the heat loss of the Earth but not the Solar heat gain. This could, it is suggested, result in an increase in ambient temperature of a degree or so and that might melt much of the icecaps. Depending upon which school of thought you belong to, this will mean either disaster by flooding or merely an increase in plant life that will compensate for the increase of energy gain.

Sulphur dioxide in the air oxidises to produce sulphuric acid.

Rainwater is made into dilute acid when it falls near an industrial chimney emitting the gas.

There are many other gases that pollute the atmosphere and waters — for more complete summaries refer to the reading list given.

Combustion also produces particulate matter ranging in size from 10 micrometres in diameter upward. Smog, smoke, haze and fog are predominantly made of particles, but not always, for optical dispersion effects can produce with gas alone the brown colours seen. Although not regarded as a pollutant in the same sense, pollen grains producing hayfever provide a similar measurement problem, for the grains are minute but powerfully annoying.

Measurement of contaminants in air, therefore, involves in the main, the determination of small quantities of gas impurities and the size and distribution of suspended particles.

POLLUTION OF WATER

In many areas of high density dwelling and industry there is a shortage of clean fresh water — 16,000 x 10^6 litres are used each day by British industry. To say fresh water is our life blood is no overstatement, for it seems all processes require it in one way or another. Power stations require immense quantities for cooling purposes, and to charge the boilers, (salt water is often used). Paper making needs it when making pulp. It takes 180 litres to produce a glass of beer, 400,000 litres to make a car. In most processes it is used only as a transport medium to wash away impurities. Such discharges are termed industrial effluents.

Nature has provided a natural purification process in water courses, and this action can handle a small amount of contamination bacteriologically.

The evaporation rain cycle is invaluable. It is, therefore, reasonable to allow a very limited amount of suitably treated effluent to go into rivers and the sea, but the natural processes must not be overloaded or the whole action ceases. However, the convenience of discharging effluent into a rapidly moving river has enticed too many people to pass their waste on to others.

The main contaminants in water come from industrial waste, sewage, and from chemicals carried from the water-shed areas by rainfall drainage. There is an identifiable water cycle, (see Fig. 3); in it the various contamination courses are interrelated.

In the 19th century, it was a sport to set light to methane discharged from some English canals! It is the absence of dissolved oxygen that is paramount in a water course, for bacteria need at least 2 ppm to convert organic carbon and nitrogen compounds into less harmful chemicals. The Biochemical Oxygen Demand (BOD) is a test designed to find the oxygen need of an effluent. It is arbitrary in nature but does provide, along with other tests such as the Chemical Oxygen Demand (COD) and Permanganate Value (PV), a measure of the degree of pollution.

Some chemicals can be most harmful, even in minute concentrations. Mercury, cadmium and lead are well-known poisons of the human metabolism, entering either through fresh water or sea-water paths. Mercury entering sea-water is concentrated in the bodies of many fish — tuna and shark have often been banned for human consumption for this reason.

Cadmium is a recently declared danger. In 1971 the reason was found why hundreds of Japanese women were suffering from bone decay leading to painful death. It was established that industrial effluent from a factory up-stream contained cadmium. This entered their bones via water irrigation used for the rice they ate.

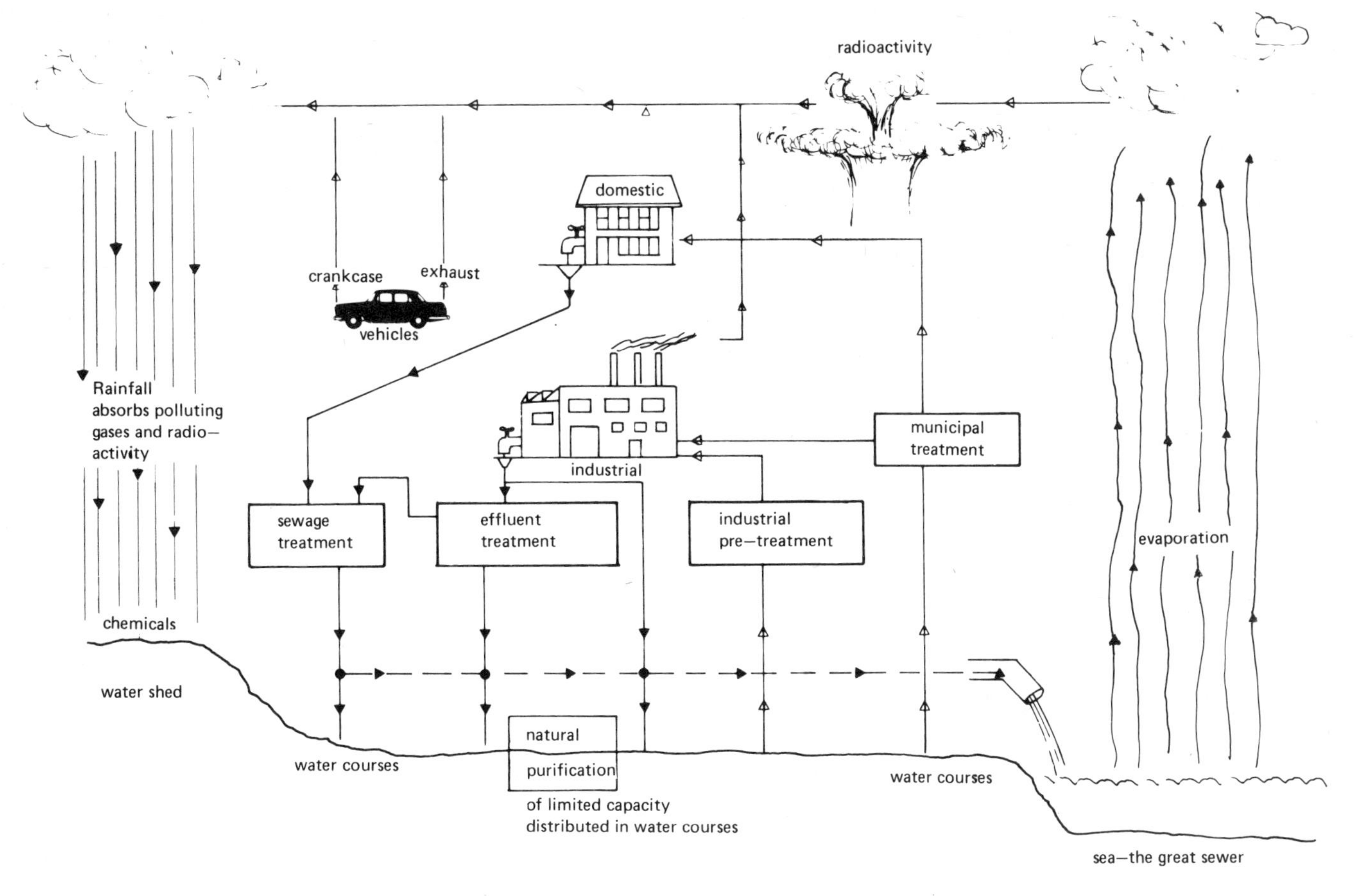

Fig. 3. Water is naturally recycled on a global basis. This basic diagram shows how pollutants enter and to some extent how they are purified.

Control of water pollution, therefore, also involves the need to measure chemical impurity levels, dissolved gas quantities and, as with air, the nature of particulates. Large solids also need consideration but their measurement is more straight forward.

Transducers needed for pollution measurement and control are, therefore, devices for measuring chemical parameters – acidity, ion concentration, specific gas content and composition, and particles. Radioactivity contamination is common to both air and water and will be covered later.

CHEMICAL ANALYSIS INSTRUMENTS

The simplest way to monitor unwanted chemical composition in a gas or liquid is to carry out conventional laboratory tests on samples. Special kits are sold to standardise the procedure. Some enable tests to be made on the spot by virtue of visible colour changes that can be matched against a chart.

Another simple way is to suspend treated paper (litmus for instance) in the fluid stream. The gas analyser in Fig. 4 is a relic used at the turn of the century to test for ammonia and sulphur dioxide in town gas.

Whilst there are cases where these inexpensive methods are satisfactory for spot checks, the need is often for a faster response and a continuous output signal that can be used to actuate control. Such instruments are almost always sophisticated and, therefore, costly. Space does not permit a complete study but those described are the commonly used instruments. Each has application in chemical analysis in general – there is nothing about chemical pollution that gives it a different need to normal analytical practices.

MASS SPECTROMETER

When a gas (which consists of atoms, or molecules made of atoms) is subjected to thermal agitation, some of it will be split into separate atoms with differing electron charges. If positively charged it is called a cation, if negative an anion. In 1907, J. J. Thompson reported a method for separating out different ions into separate locations where an individual measure of each can be made. This instrument, called a mass spectrometer is shown diagrammatically in Fig. 5. It can be used to monitor gas composition as a continuous process.

The example chosen is used in the iron and steel industry to monitor – on line – waste gas composition from blast, oxygen and electric-arc furnaces.

The gases to be studied are sampled

Fig. 4. This elegant coal-gas monitor of the 1900's has two treated papers hanging in front of the gas stream. The sampled gas passes through to be burnt at the top.

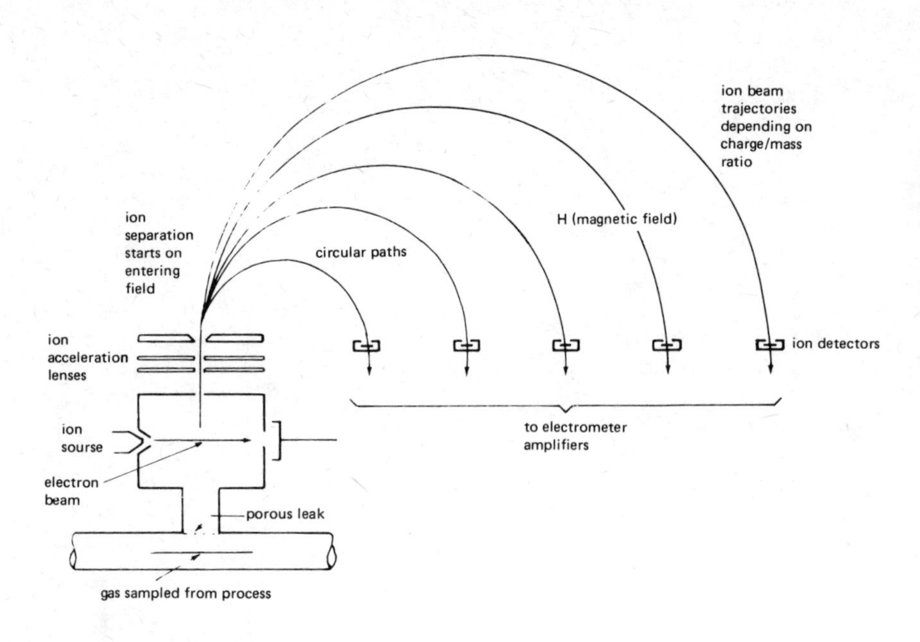

Fig. 5. In the mass spectrometer, ionised gas atoms and molecules are separated into groups depending on their electron-charge/mass ratio.

by bleeding some from the process. They are then pumped past the ionising chamber. An ion source ionises the gas mixture and ions are accelerated by electron lenses to follow the same initial path. Upon entering the magnetic field zone they take up circular paths of radius depending upon their charge/mass ratio. Finally they impinge on ion detectors that produce current proportional to charge. In this way the gas is separated into its constituent parts and a spectrum is formed in space (hence the name spectrometer). Detectors, if placed correctly, will sense specific ions only. Process control mass spectrometers are available commercially.

A chance of ambiguity exists, for there are a number of combinations of charges and masses that produce the same charge/mass ratio even though the element is different. Other tests may be necessary to reduce the risk of error. This is a common difficulty with most analysers, especially when the number of constituent gases rises.

The mass spectrometer is extremely versatile being able to detect any chemical substance that can be ionised. In the steel works example, it is used to measure nitrogen, carbon monoxide, hydrogen, oxygen, carbon dioxide, and water vapour. Other uses have been to analyse the smell of the land after rainfall, the odour of packed apples – it is the best general purpose sniffer available, (the biological sense of smell is however, more sensitive). Well designed instruments can distinguish separate ions having charge/mass ratios differing by as little as one part in 10,000.

OPTICAL SPECTROMETERS

In this spectrum-based measuring device, it is the dispersion of electro-magnetic radiation into the various wavelengths (colours if in the visible region) that is used, not the deflection of ions. A sample of gas under study is heated or excited by forming an electric discharge, as shown in Fig. 6. A collimator unit provides an essentially parallel beam of the resulting radiation, as though the source were at infinity. This beam is dispersed into its 'colours' which spread out around the output area. Dispersion is achieved with either a prism, or as in many instruments, with a grating having ruled surface grooves at a pitch appropriate to the wavelengths of interest. Gratings of both transmission and reflection type are used. The output optics see a defined field of view that can be observed manually or with suitably placed photo-detectors (ranging from relatively insensitive photo-cells to photo-multipliers). The intensity of the radiation seen at the various angular positions provides a unique set of data for a given gas. Rotation is usually achieved by slowly scanning the dispersion element keeping the output stage fixed. Spectrographs using photo-detectors are known as spectrophotometers. Spectra (the radiation bands and lines) produced by a source including the gas to be analysed are obtained as emission in this type of spectrometer.

Black body radiation (see the earlier discussion on temperature) produces an emission spectrum that is continuously graded from colour to colour. In contrast, radiation from gases contains one or more sharply defined lines at precisely known wavelengths. This is explained by quantum theory which shows that energy will be emitted at certain wavelengths only. Knowledge of the prism or grating and the geometry of the instrument enables the line set for particular gases to be determined and hence the analysis of the sample placed in the source.

Many adaptions of the spectrometer principle exist. In the spectro-photofluorometer shown in Fig. 7, identification of chemical compositions is by virtue of fluorescent and phosphorescent characteristics of compounds. The molecules of the sample are excited by ultraviolet or visible radiation producing luminescence that radiates at longer wavelengths – the energy is transformed in wavelength. It is claimed that the sensitivity of this fluorometry technique exceeds normal spectrophotometry by several thousand times: parts in 10^{12} sensitivity is often obtained. A lot depends upon the substance being analysed, of course.

In the absorption technique, use is made of the property of a gas to absorb radiation, an effect that depends upon the wavelength of the radiation supplied and composition of the sample. The atomic absorption flame spectrophotometer shown in Fig. 8. became a generally accepted reality in the 1960's after a decade of research at the CSIRO. Originally it was considered that emission spectra monitoring was the better way because the gas produced large amplitude signals. Overall, however, absorption monitoring is superior. The gas (or liquid) to be analysed is fed into a flame, through which radiation from special spectral-line lamps is passed. Study of the spectrum of the energy leaving the flame provides wavelength – amplitude relationships that are again unique to each gas. The reason for the superiority of absorption is that the source, being spectrally pure, enables a better overall signal-to-noise ratio to be obtained – in the emission method the detection signals include many unwanted emission lines that cannot be eliminated in the same way.

In spectrometers each gas is defined by its lines and their positions. Often they are not sharply defined and, further, the spectra may be very similar. One way to increase the certainty of resolution is to feed the scan signal obtained into a powerful digital computer and use correlation techniques (described earlier in flow measurement) to test the unknown with a standard spectrum. The use of a computer is, however, expensive. A more economical method uses a mask in the spectrometer exit slit that has a transmission-versus-position characteristic of the standard spectrum being sought. The unknown spectrum is vibrated across the slit and the total transmission of energy through both the mask and the exit slit is an optical correlation of the two. The amplitude of the signal with position of the mask provides a test of the match of the two spectra. The set-up used in a production correlation spectrophotometer is shown in Fig. 9.

The non-dispersive infrared analyser, NDIR for short, is commonly used in air pollution measurements. Its principle is based on the selective absorption of gases but no dispersive element is involved, that is, no spectrum is formed. A heated wire provides broad-band infrared energy which is split into two identical power beams, see Fig. 10. They are mechanically shuttered to first pass one beam, then the other at about 1 Hz frequency. In the reference path, a transparent cell is inserted that is filled with a known non-absorbing gas at the infrared wavelengths provided. In the other path the cell is filled with the sample gas. Both beams then impinge onto a common detector cell, also filled with gas. If the sample cell contains gas that absorbs energy the detector cell will be heated slightly less in one half of the cycle than in the other. This produces a cyclic heating effect that manifests itself as pressure changes in the detector. A microdisplacement transducer — capacitance perhaps, monitors the minute vibrations of a diaphragm mounted on the cell. Synchronous detection, derived from the chopper supply, enhances the signal-to-noise ratio. The fluctuations are rectified and converted to dc indicating the degree of absorption as the amplitude of the final output signal. Filter cells are used to reduce the risk of ambiguous operation by removing unwanted wavelengths before the radiation enters the sampling cell. The method is fast to respond having a response time of the order of seconds.

An NDIR instrument can detect carbon dioxide down to concentrations of 10 ppm but the presence of carbon monoxide, water or methane can introduce considerable error. It can also be used to detect sulphur dioxide down to 2 ppm but again if water and carbon dioxide are present the results are invalid.

The principle used can also be worked in the ultraviolet range of the

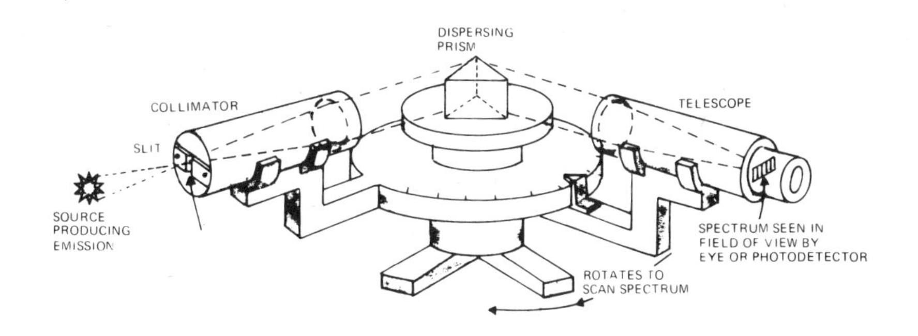

Fig. 6. In the optical spectrometer radiation from a gaseous source is dispersed to form a spectrum of lines and bands unique to each element.

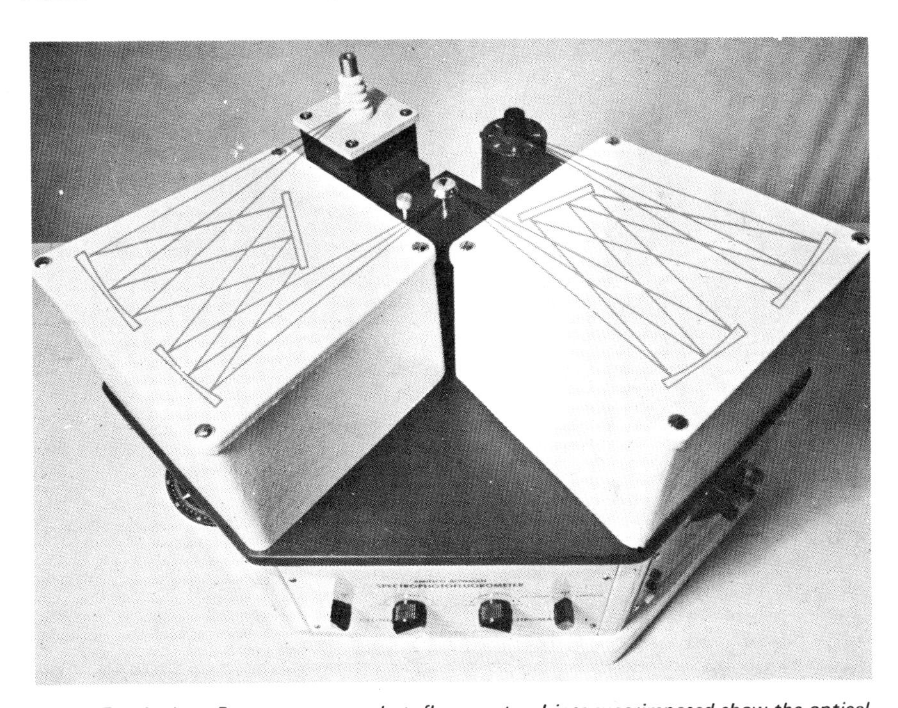

Fig. 7. The Aminco-Bowman spectrophotofluorometer. Lines superimposed show the optical paths of the excitation grating monochromator (on the left) that illuminates the sample placed in the centre and the emission dispersing grating monochromator (on the right) that is used to analyse the spectrum of the luminescence. The source is a Xenon lamp and the detector, a photo-multiplier of appropriate wavelength sensitivity.

Fig. 8. The Australian designed atomic-absorption flame spectrophotometer.

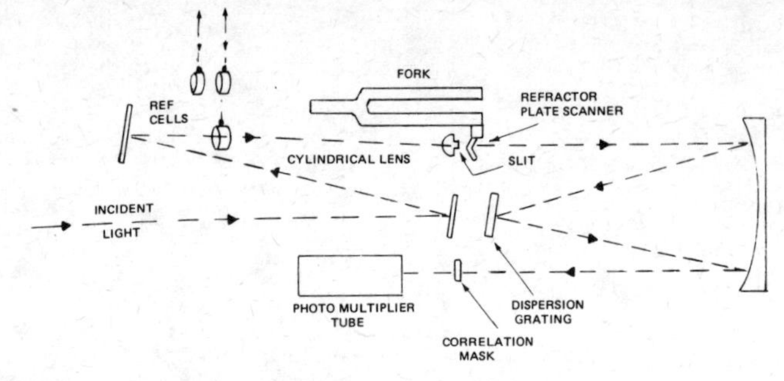

Fig. 9. Use of a mask to perform optical correlation of the spectrum of a vapour.

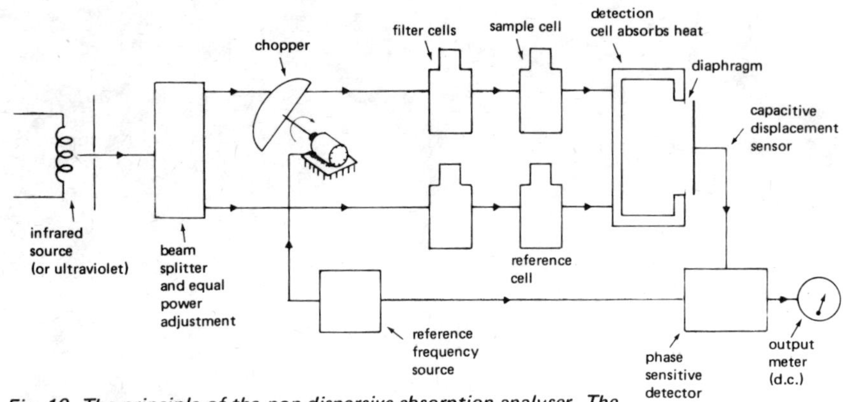

Fig. 10. The principle of the non-dispersive absorption analyser. The source is either in the infrared (the NDIR instrument) or in the ultraviolet region.

spectrum if a suitable source (tungsten lamp) and detector are used. Its main air pollution application has been for the detection of nitrogen dioxide from 5 ppm to 200 ppm concentrations.

FURTHER READING

"Threats to the environment: a world wide review" in Chambers Year Book 1972, 120-124, International Learning Systems Corporation 1972, London.

"Air pollution" A. Gilpin, Univ. Queensland Press, St. Lucia, 1971 (An international coverage).

Specification for gas analysers" A. Verdin, *Trans. Inst. Meas. Control,* London, 1971, 4, 44-46.

"Continuous gas analysers ..." C. C. Simpson, *Inst. of Instrumentation and Control.* Australia. Symposium on the "Advance of process instrumentation" Sydney. 1971.

"A Survey of measurement and control of pollutants" D. J. Garrod, *Trans. Inst. Meas. Control,* London, 1971, 4, 253-262.

"Where there's muck there's brass" A. Conway, *New Scientist,* 1972, 54, 376-378.

"Recent developments in continuous analytical instrumentation for measurement and control of pollutants". L. E. Maley, *Trans. Inst. Meas. Control,* London, 1972, 5, 3-8. (Discusses correlation spectrophotometry).

"The unclean sky" L. J. Bathan, Anchor Books New York, 1966.

"Fundamentals of Analytical Chemistry". L. Pataki and E. Zapp, Adam Hilger, London, 1977.

"Spectroscopy and its Instrumentation". P. Bousquet, Adam Hilger, London, 1971.

"Analysis Instrumentation", Vol. 15. J. F. Combs, F. D. Martin and W. H. Wagner, *Instrum. Soc. Am.,* 1977.

"Non-dispersive Infra-red Gas Analysis in Science, Medicine and Industry". D. W. Hill, Plenum, New York, 1968.

"Gas Effluent Analysis". W. Lodding, Marcel Dekker, New York, 1967.

"Instrumental Analysis for Water Pollution Control". K. H. Mancy, Ann Arbor Science, *Ann Arbor,* 1973.

CHAPTER 12

POLLUTION MONITORS — 2

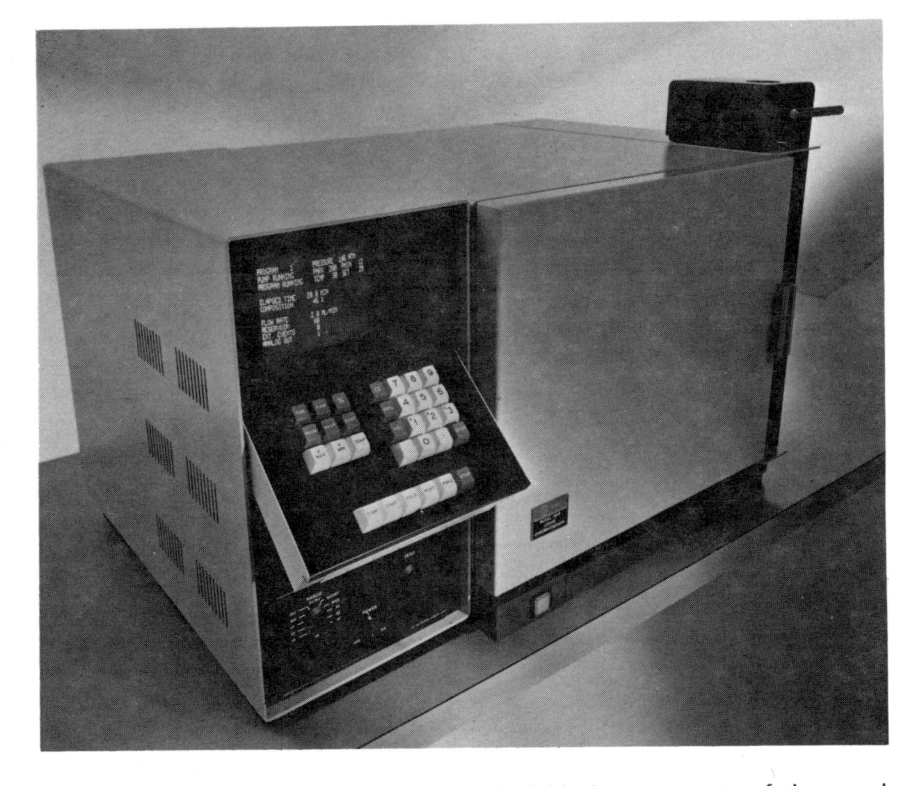

Keyboard control and CRT display are incorporated into this Varian liquid chromatograph to simplify the operation whilst giving greater capability to the operator.

Pollution of water and air occurs in distinctive groups each requiring different measurement approaches. These groups are unwanted chemicals, particulate matter and radioactivity.

CHROMATOGRAPHY

In the analytical methods described previously, the various chemical constituents of a gas or liquid were identified by separating each, either directly (as in the mass spectrometer), or indirectly, (using the spectrum of radiation). They were then sensed at the different spatial locations.

Chromatography is another procedure by which the chemicals are initially separated in some way so that each may be identified. When a sample mixture, such as a gas or liquid, is passed through, or over, surfaces of another material of different chemical phase (for example, as gas passing over a solid) the transmission times of the individual components of the sample are selectively delayed. They emerge through the column (of different phase material) in a specific time sequence.

In gas chromatography the gas to be analysed is either percolated through a porous solid column (charcoal, silica gel are used) or over a large-area liquid film. The former is known as a gas-solid chromatography, GSC for short, the latter GLC. Other methods used include liquid-solid and liquid-liquid systems. Here, only gas chromatography will be discussed as this illustrates the general principles.

Chromatography had its origins in the mid 19th century. It really became established around 1905 when Ramsey devised a method to separate gases and vapour mixtures, and Tswett used the principle to extract chlorophyll from plant pigments. The latter biochemist coined the name now used because of the coloured bands he obtained down a vertical calcium carbonate column. Chromatography is formed from the Greek words for colour and write. To prevent possible confusion it must be made clear that colour is rarely a parameter in modern chromatography.

The basic essentials of a gas chromatograph (established by James and Martin in 1952) are shown in Fig. 1. An inert carrier gas passes through the separation column to a detector cell. The unknown gas sample is injected into the inert gas carrier flow prior to its entry into the column. The various constituents of the gas arrive at the detector at different times, producing peaks on the recorder chart as the paper moves with time. The sharpness of the peaks, their amplitude and relative time positions identify the sample. It is essential to hold the gas and column at a steady temperature; commercial units enclose the critical areas in a temperature-controlled oven held to 0.1°C limits. Higher than ambient temperatures also enable liquids to be vaporized and treated as gases.

Some components are strongly retained by the column, emerging only after a considerable duration. To speed up the process the temperature is

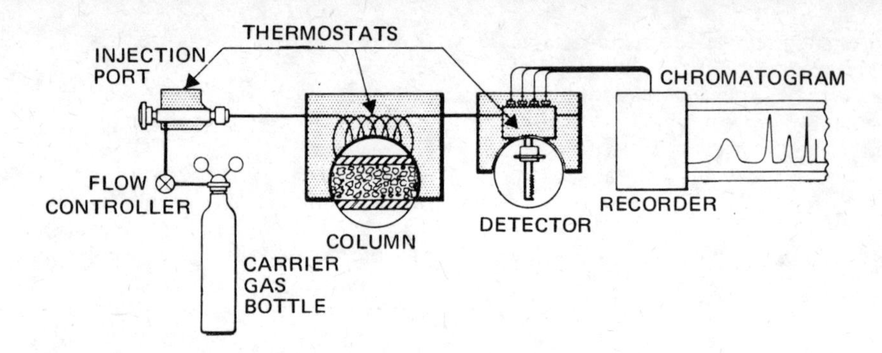

Fig. 1. Essential components of a (Varian) gas chromatograph.

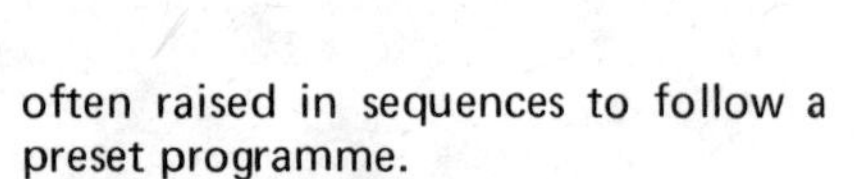

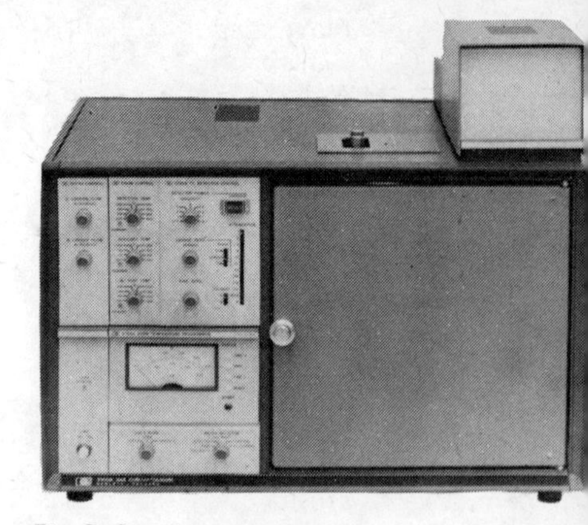

Fig. 2. Gas chromatograph marketed by Hewlett-Packard (series 5700)

often raised in sequences to follow a preset programme.

Detection sensitivity depends upon the detector used to monitor the emerging gases; it ranges from parts per thousand to parts per billion. To quote a Varian example, one form of detector can sense certain chemicals down to a molecule of sample in every 10^{10} molecules of carrier gas. Such sensitivity has enabled the method to be used in the analysis of odours in foodstuffs. Units such as the one shown in Fig. 2 are moderately expensive, but less versatile, cheaper, units are available.

Detectors in use are varied and numerous, the main two being the ionization detector and the thermal conductivity cell.

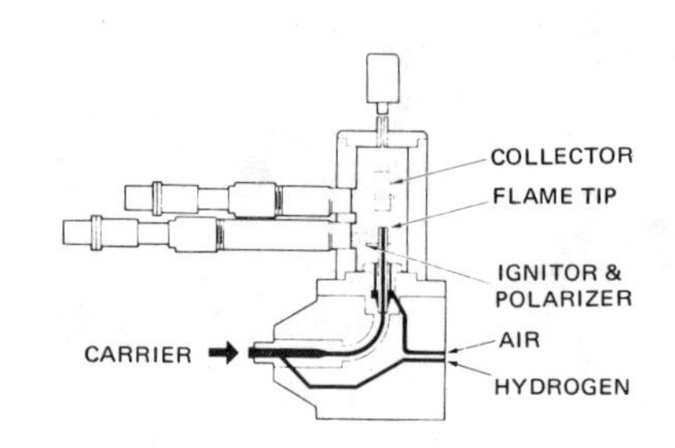

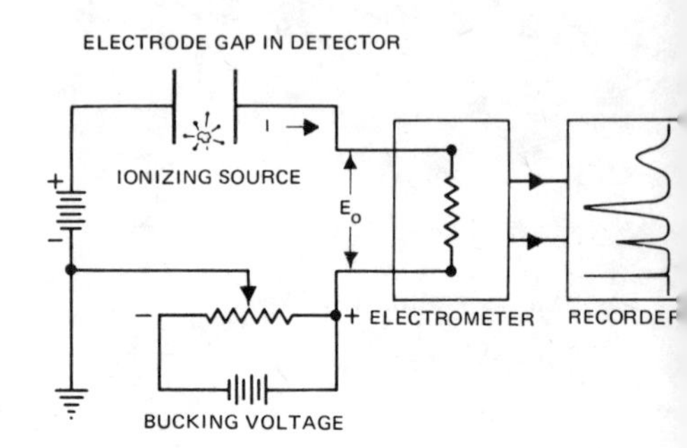

Fig. 3. Cross-section of Varian flame ionization detector and typical electronic circuit used to detect the ionization.

FLAME IONIZATION DETECTOR

When a carrier gas of hydrogen is burned it produces a colourless flame. Organic compounds cause it to burn yellow with a height and luminosity proportional to the amount of hydrocarbons present. Flames produce ionized gases in such cases and this effect is used to obtain a more accurate measure of the arrival events out of the column preceding the detector. These cells are called flame ionization detectors (FID). For reasons not fully understood, organic compounds ionize in a flame, and suitably placed electrodes (Fig. 3), detect the minute current flowing. High input impedance amplifiers are needed because the flame resistance is around 10^{12} ohms. Advantages of the FID are that it does not detect water vapour or air, is simple and has a wide response range. These characteristics make it particularly suited for pollution measurements of water and air.

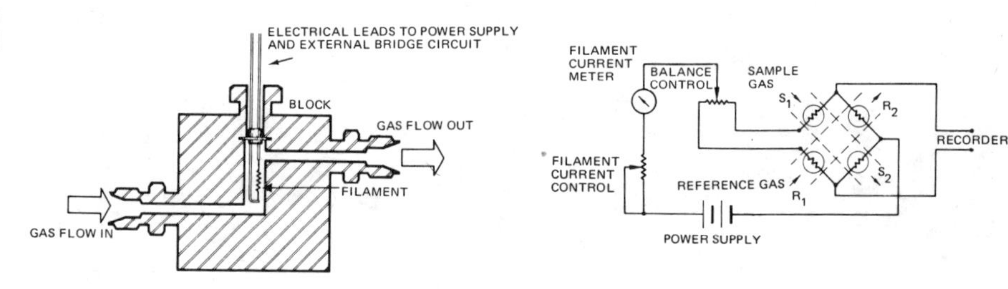

Fig. 4. Schematic diagram of one element of a thermal conductivity detector. The sample gas is made to flow across two resistances, the reference gas across the opposite pair.

THERMAL CONDUCTIVITY CELL

This detector, introduced by Claesson in 1946, is also commonly employed in chromatographs. It operates by measuring the thermal conductivity of the gas. A heated filament, suspended in the flow, will vary in temperature as the heat is conducted away by the changing conductivity gases emerging from the column, thus changing its resistance. (Very similar in operation to the hot-wire anemometers used to measure flow rates). These are also called katharometers or simply TC units. A schematic of a TC cell is shown in Fig. 4 together with the layout of a typical electrical arrangement. Note that the reference gas passing into the column before injection of the sample is fed across two detector filaments of the bridge and that the outlet gases (carrier plus separated constituent) pass over the other two. This technique makes best use of the properties of a bridge circuit to eliminate unwanted common signal effects existing in the apparatus.

With thermal conductivity cells the gas flow limits the temperature rise of

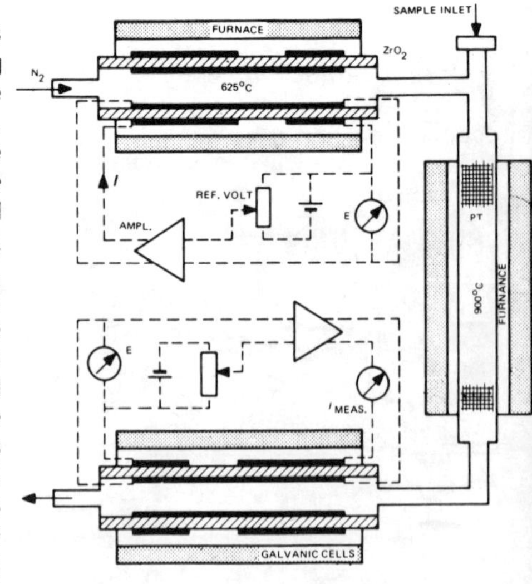

Fig. 5. Layout of a chemical oxygen demand detector using high-temperature galvanic cells.

the filaments. Flow is essential, when the detector is energized, to prevent burnouts. Thermistor sensors are sometimes used instead of the tungsten wires.

Flame ionization and thermal conductivity detectors are the more common types used, but others exist that might be more suited. They include electron capture cells for detecting alkyl halides, carbonyls, nitrides and nitrates — but not hydrocarbons (useful for pesticides). The helium detector may be used for extremely sensitive analysis of all compounds, provided they are pure enough to begin with; the alkali-flame detector for sensing phosphorus compounds (the newer forms of cides that have largely replaced the now unpopular hydrocarbon forms); and the gas-density balance for the analysis of corrosive compounds. Space does not permit descriptions; they are to be found in the listed texts. Where mixture separation is not needed the column can be discarded, passing the gas through the detector only. Several specific analytical instruments operate this way.

The similarity between the amplitude-time recorder plots from a chromatograph and a spectrograph is striking and the use of correlation techniques appears relevant in the detection process of chromatography (correlation was encountered earlier in the discussion of flow-meters). To date, however, there appears to be little gain when the extra difficulties are accounted for. A study made in 1968 (by Davies) showed that there were two main drawbacks. Firstly, extra gas sample was needed causing the column to operate in a non-linear mode and, secondly, the correlation process was expensive. Since 1968 the latter objection has been lessened by the introduction of commercial units. Even so, a study by Moss and Godfrey in late 1972, concluded that the case is still not strong but might expand in pollution measurements where specific equipments could be marketed thus cutting the cost.

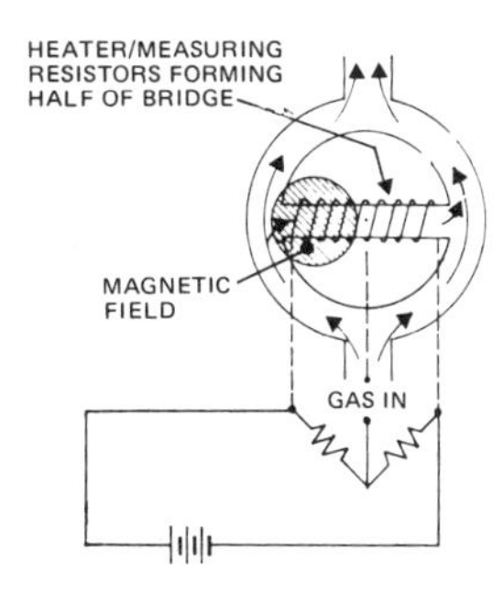

Fig. 6. The Harmann and Braun magnetic wind oxygen analyser now used widely by many manufacturers.

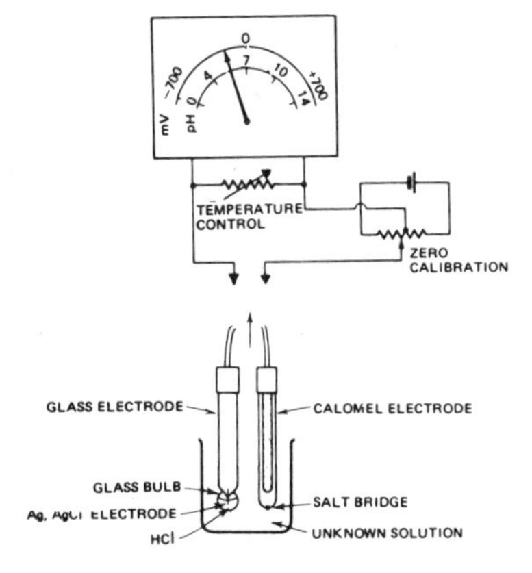

Fig. 7. pH meter using glass and calomel electrodes.

DETECTION OF OXYGEN

Detecting oxygen levels in air, water and industrial processes is a commonly needed measurement. This has led to the development of a number of specific oxygen detectors.

By removing the oxygen (with absorbent columns) from a known volume of gas, and remeasuring the volume, it is possible to determine the oxygen content. This is an old established method but as it does not supply a continuous electrical signal the method has only limited use.

Oxygen analysers exist for use in continuous processes and are mainly of two types; those using electro-chemical principles and those making use of the magnetic properties of oxygen. In principle, one form of the first is based upon a special cell in which oxygen concentration is controlled by an input voltage.

A schematic diagram of a Philips unit devised to monitor the COD (chemical oxygen demand) of possibly polluted water is given in Fig. 5. The special cell consists of a zirconium oxide tube having porous platinum electrodes attached. When hot (the reason for the oven at 625°C) the tube develops a voltage between the electrodes that is related to the partial pressure of oxygen on each side of the tube. A current passed through the cell wall transports oxygen through the wall. With electronic feedback the oxygen partial pressure of an unknown gas can be compared with a known gas. In the COD measurement two such cells are used. The upper provides a constant concentration (p.p.m.) of oxygen in a nitrogen carrier. This enters, along with a minute sample of water to be tested, a furnace at 900°C which oxidizes and removes all oxygen. The gas then enters a second cell where the oxygen demand is met by the electric-servo oxygen transporter. The difference between this requirement and the original concentration is a measure of the COD of the liquid. The method can measure COD values ranging from 1 to 5000 mg of oxygen per litre in just two minutes. Such detectors are termed high-temperature galvanic cells, and are specifically sensitive to oxygen, so water and carbon dioxide do not upset the analysis. Combustible pollutants, however, may consume more oxygen in the furnace indicating a false COD value.

The polarographic electro-chemical oxygen method, so called because the rate is controlled by the electrode area, uses oxygen diffusion through a Teflon membrane at ambient temperatures to produce a microampere current between two separated electrodes (26.3 μamps/p.p.m. in theory) — the Mackereth cell has a lead anode inside a silver porous cathode, the two having an electrolyte between them. Polarographic electrodes can be made as small as 2 mm in diameter.

The second type of oxygen detector operates on a quite different principle — the paramagnetic properties of the oxygen molecule are used. In the O_2 molecule two electrons are unpaired providing a strongly paramagnetic condition. Faraday discovered this in 1848, but it was not until the 1940s that an oxygen detector was produced using the principle.

The original magnetic detector used an effect known as magnetic wind. Referring to Fig. 6 the incoming gas containing oxygen, parts to both sides with some entering the cross tube. Because of the intense magnetic field, oxygen in the tube is attracted to one side. The heater raises its temperature reducing the magnetic property of the oxygen thus pumping it out; flow of oxygen results across the entire tube and this is detected by monitoring the resistance of the heater winding. Error can occur if the carrier gas is not constant in purity, for this will alter the heat-loss of the filament. The cross tube should also be horizontal otherwise gravity flow will occur. Hydrocarbons upset the method considerably. It is sometimes called a thermal magnetic analyser. In the more advanced Quincke analyser most of these defects are eliminated — at the expense of requiring a continuous supply of nitrogen.

In 1954, Linus Pauling devised another magnetic method that is less prone to errors caused by hydrocarbons. In his detector, two diamagnetic glass spheres, mounted to form a dumbbell, are suspended on a torsional suspension inside a measuring cell. A non-uniform magnetic field is

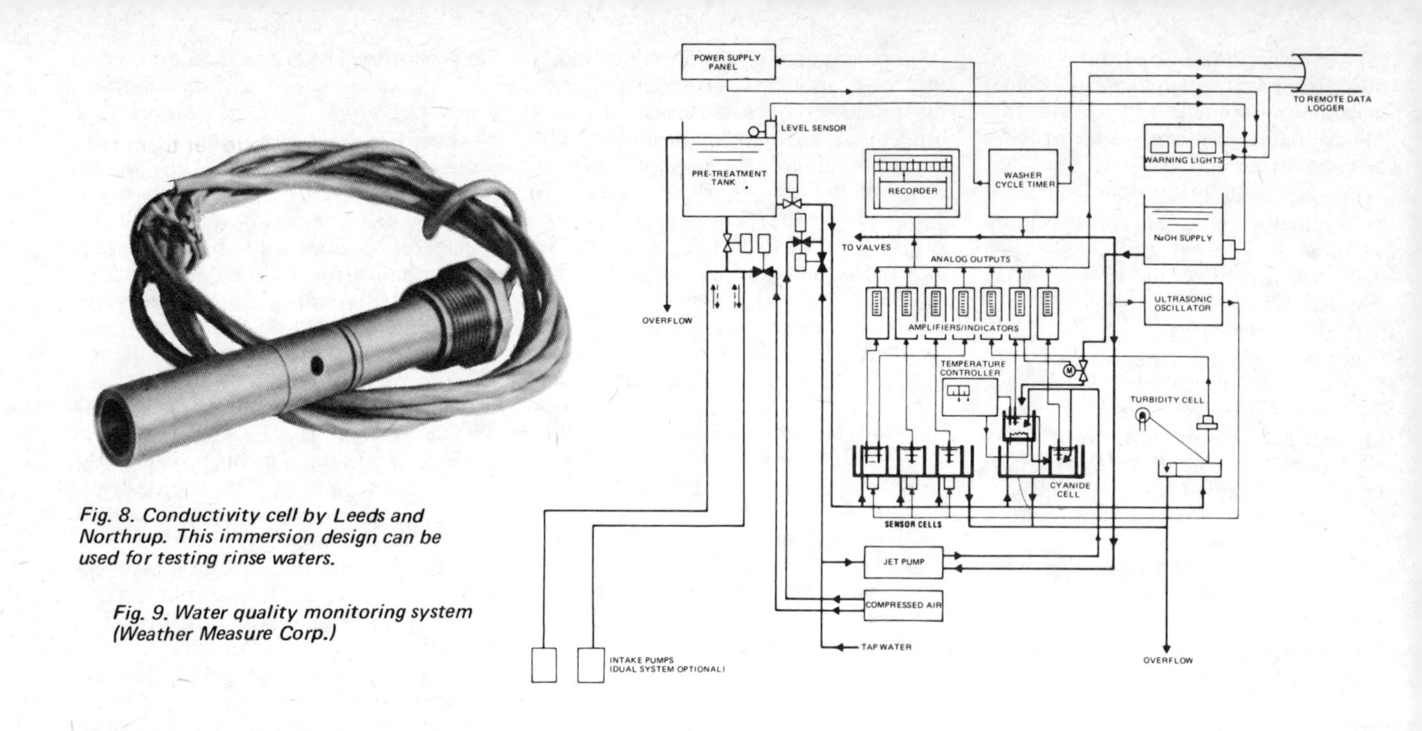

Fig. 8. Conductivity cell by Leeds and Northrup. This immersion design can be used for testing rinse waters.

Fig. 9. Water quality monitoring system (Weather Measure Corp.)

applied across the cell causing the dumbbell to rotate to an equilibrium position. Changes in oxygen level in the cell alter the field, causing the beam to rotate. Movement is sensed by a microdisplacement transducer. Suspensions are made of quartz or platinum fibres. More advanced cells of this type use the force-balance technique to restore the beam to a null-position. Many gases are paramagnetic, but oxygen is only approached in magnitude by nitric oxide and nitrogen dioxide; other gases of interest being considerably less paramagnetic.

ELECTROCHEMICAL MEASUREMENTS

Two plates suspended in a liquid form a primary cell and a voltage occurs between them that depends upon the plate materials used and the liquid composition. This concept can be used in many ways to arrive at the impurity level of the solution. It can be used, firstly, as a battery, measuring the emf with no current flow (potentiometric analysis); as an electrolysis (or coulometric) cell in which current flows consuming energy; or as a resistivity (or conductivity) cell.

In potentiometric analysis, two half cells must always be used, for the voltage of a single plate to liquid half-cell is not meaningful. Quoted electro-potentials are referred against a standard cell to obtain a working calibrate arrangement, the standard hydrogen electrode (SHE) being the arbitrary value assigned for such comparisons. The SHE is not, however, entirely practical and other reference half-cells such as the saturated calomel and silver-silver chloride electrodes are used instead, having first been calibrated against the SHE. Half-cells are connected to the liquid with salt bridges to enable ions to transfer without diffusion of the electrolytes needed in each half cell.

A common potentiometric measurement is that of pH, the measure of free hydrogen in concentration in a liquid – the degree of acidity or alkalinity. The observed potential of a cell-pair, less that of the reference cell at 25°C, equals 0.05195 times the pH value, the number coming from a simplified form of the Nernst equation explaining the electro chemical process. So called glass and calomel electrodes are used together in pH determinations as shown in Fig. 7. In the calomel electrode a saturated solution of mercurous chloride (calomel) and potassium chloride is placed over a mercury layer electrode. A salt bridge enables the ions to flow. The glass electrode has a silver wire dipping into a hydrochloridic solution. This is contained inside a glass bulb that acts as a membrane separating the acid from the sample solution, as well as forming a container. Ions migrate through the glass but as the resistance of the membrane is typically 30 megohm a relatively expensive readout amplifier is needed.

In pH meters, such as that shown diagrammatically in Fig. 7, the electrode pair operate a high input impedance millivolt meter needing a scale of ±700 mV to cover the 0-14 pH range.

Compensation for temperature is essential, for the 0.059 constant is correct only at 25°C. Other electrodes available are the quinhydrone electrode useful in bio-chemical analysis, the platinum electrode that is non-corrosive, but reads incorrectly in circumstances where chloride ions exist, the mercury electrode suited for chromium potential measurements and bimetallic electrodes made of platinum and palladium or tungsten. Operation of the latter is not completely understood.

In the electrolysis or coulometric analysis, current is made to flow either at a constant value or with a constant applied voltage. Flow is established when the voltage applied exceeds the normal (back emf) cell voltage. For example, a platinum plate and a copper plate in a solution of sulphuric acid has a back emf of 0.87 V. Faraday's law states that 96,500 coulombs (a coulomb is an amp per second) of electricity are needed for each equivalent of a chemical reaction. Hence the amount of current consumed enables the substance to be analysed quantitatively. The method is easily automated and is popular for long term analyses.

Conductometry is the third electrochemical method, and, as the name implies, relies on measurement of the specific resistance of the liquid. Cells can be made of glass having platinum electrodes but more modern designs like that shown in Fig. 8 are made of high-impact strength non-corrosive plastics such as polyvinyl dichloride, PVDC, with embedded gold-plated nickel or platinum electrodes. The fluid is either made to flow through the cell or the cell is simply immersed in the sample. Alternating current bridges are usually used, operating at 1-10 kHz. Ten MHz units have been marketed under the name Oscillometers. For dc operation, non-polarizing electrodes such as silver/silver chloride might be suitable.

Each electrochemical method can be used to monitor water quality but a number of detector cells are needed if all pollutants of interest are to be monitored. Commercial multi-sensor monitoring consoles exist – Fig. 9 is a block diagram of a versatile unit that will continuously monitor pH, conductivity, dissolved oxygen (DO), turbidity and numerous specific ion concentrations (bromine, chlorine, sodium, cadmium, iodine, cyanide, etc.). Sensing electrodes are automatically cleaned at regular intervals by ultrasonic vibration.

This outline is, by necessity, a brief resume of the chemical analytical instruments used commonly in water and air pollution measurements. Two other powerful analytical techniques, nuclear magnetic resonance (NMR for short) and neutron activation analysis, are applicable but are not used as extensively in routine pollution measurements, being limited by cost or transport factors. They are, nevertheless, worth considering. Details can be obtained in the suggested reading.

PARTICLE MONITORS

The presence of particles suspended in air or water may present a health hazard or impair visibility to such an extent that the air or water is polluted.

Fog, haze, mist, smog, call it what you may, can be the result of optical dispersion or of suspended particles, ranging in size from smokes with 0.1 μm diameters to grits of 100 μm. Smoke, airborne bacteria and fine fibres are in the 1 μm size range, fine dusts from 1-20μm and coarse dusts 20-80 μm. Devices for measuring the concentration of particles are known as turbidity sensors (in water) or nephelometers (Greek for cloud) in air.

Particles may be permanently suspended by virtue of their small size compared with the molecules of the medium or may be transiently suspended by virtue of an upward velocity, for example, as found in chimney stacks. Coal and oil furnaces are the worst offenders in industrial areas, with cars adding considerably by emitting unburned hydrocarbon particles.

Average particulate concentrations in remote non-urban areas of the United States lie around 10 $\mu g/m^3$; in urban areas around 100 $\mu g/m^3$. The heavily polluted areas go as high as 2 mg/m^3. An accepted safe level of particle precipitation is around 200 $mg/m^2/day$ (15.4 $tons/mile^2/month$). Brisbane City suburban records for 1969, indicated values of 7-35 $tons/mile^2/month$ indicating that some suburbs were unhealthily polluted in this way. This amount of dust is easy to produce! A 200 MW coal-burning power station operating with only 0.7 percent dust loss from the chimneys would pour out 20 tons of dust a day. In the 1950s, records for the Pittsburgh area in the United States ran as high as 2 $g/m^2/day$ (170 $tons/mile^2/month$).

The cheapest method to monitor particle fallout rates is to let them fall for a given time onto known size slides or plates which are later examined by counting the particles, using a microscope; or weighing the carrier before and after. Fans or suction are used to increase the yield.

In the airborne bacteria sampler shown in Fig. 10, a culture plate, surfaced with a nutrient solution, is slowly rotated under the dome cover. Air is drawn in by a low-vacuum pump, passing through a slit positioned above the rotating plate. Bacteria come to rest on the plate and a colony begins to grow. After the sample period is complete the plate is removed and incubated. The record obtained of the plates is also shown in Fig. 10. Up to position three the bacteria were freely moving in the air. At three an ultraviolet lamp was turned on – the record shows the diminution of cultures after the event.

OPTICAL METHODS FOR MEASURING TURBIDITY

The most direct method is to monitor the loss of illumination intensity of an optical beam radiating through the smoke or haze. An installation devised by staff of the CERL (Central Electricity Research Laboratories) in Britain is shown in Fig. 11. Note the Everclean windows that help to overcome signal loss common to viewing windows in such dirty conditions. Air is pumped into the sampling tube at five second intervals to purge the system clean and reset the zero.

Fig. 10. Casella airborne bacteria sampler MKII and record produced.

Aircraft runways can become clouded and when this happens the pilots desire a measure of the degree of visibility. The Transmissometer is the instrument becoming accepted to perform this task, displacing personnel who make subjective assessments of visibility. In the Transmissometer, a powerful beam of light, often a spark discharge pulse source, is transmitted to a receiver. A telescope gathers the radiation arriving, directing it to a photo-tube or photomultiplier detector. The response of the detector is made to match that of the eye in visual transmission testing. There is an increasing use of this principle on freeways where fog is encountered. Another form of the same concept has the receiver mounted at the detector; back reflected light is used to determine the visible range. Visibility meters can operate over ranges from a hundred metres to 25 km.

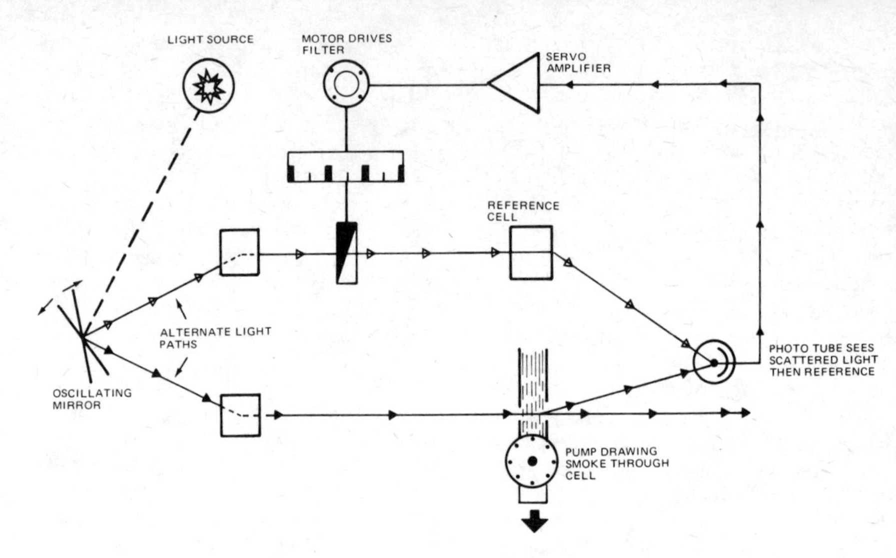

Fig. 12. Sigrist dust monitor operates from the forward scattered stray light produced by a light beam passing a smoke. It can detect concentrations as little as 0.005 mg/m³.

In practice, sophistication is needed to eliminate various sources of error. Firstly, it is desirable to modulate the light to overcome the effect of ambient light. Secondly, a portion of the outgoing light is referred back to the incoming to reduce the influence of source intensity variations. Another feature often incorporated, uses the same detector to sense the outgoing and then the returned beam thus eliminating differences in photocell characteristics. The null-balance technique is shown in Fig. 12. The filter wedge attenuator is servo-controlled to obtain a balanced photocell output from each of the two paths as the mirror is rocked from side to side at 600 Hz. The sample cell is compared against a reference until a null is achieved – the position of the optical attenuator is then a measure of turbidity. In some designs light scattered at 90° to the beam is used, for this reduces the errors due to colour or shape of the particles. The turbidity of solutions can be determined in a similar manner, the solution being placed in a test tube that is placed between the transmitter and the detector.

When the particles are large it is the settling rate that is of interest. The CERL dust monitor, as shown in Fig. 13, operates on the principle that the heavy dust will fall out of the flow onto a glass collector plate reducing the transmission. Again, air is used periodically to blast the windows clean.

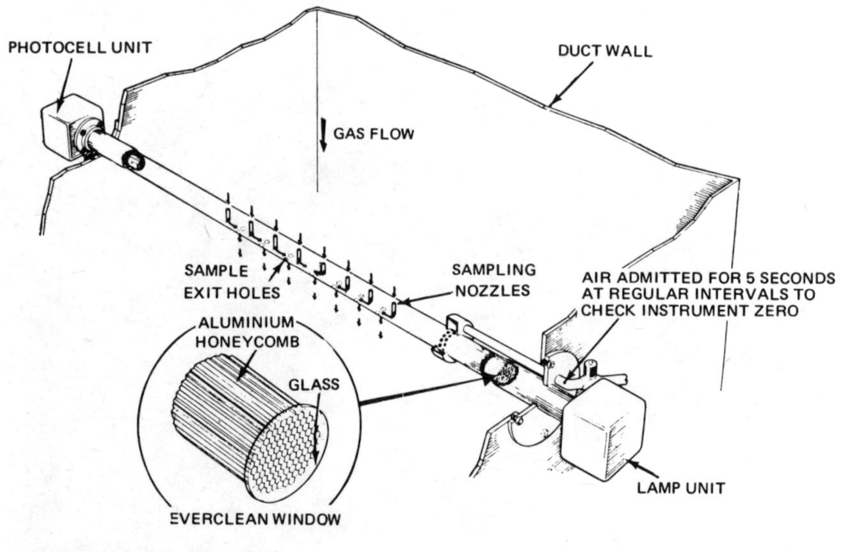

Fig. 11. Smoke density recorder designed at the Central Electricity Research Laboratories CERL in Britain. The patented Everclean windows use a long, thin aluminium honeycomb to prevent the formation of particles on the glass.

PARTICLE COUNTING

An interesting method marketed by Particle Data Inc. makes use of the change in resistance of liquid flowing between electrodes as particles flow in suspension. The particles are first added to a suitable electrolyte that is then drawn steadily through an orifice (with electrodes) that detects resistance changes. The output pulses are amplified and then integrated or distribution analysed into size–time charts. Ranges covered go from 0.3μm to 300 μm. Flow rate is regulated to reduce coincident occurrences of the particles. As in most nephelometers output is given as a logarithmic scale. Special data processing equipment is available to perform the distribution analysis.

Other non-optical methods include measuring the charge removed from electrodes as the dust passes, and charge carrier rates between electrodes.

When the particles become very large, as in sewage and slurries, they can be detected by capacitance or electromagnetic changes. Certain flow meters (see previously) operating on this principle can yield data on particle size whilst acting as flow sensors.

RADIOACTIVITY

Corpuscular radioactive radiation occurring naturally and synthetically emits packets of energy as alpha, beta and gamma rays. These, and X-rays, lie in the electromagnetic radiation spectrum above 10^{17} Hz. Such radiations can be most harmful, especially when it is considered that small doses go undetected only producing symptoms years or generations later. Nuclear radiations have the property of decreasing in radiation strength according to an exponential law. The rate of loss of activity is conveniently described by the time taken to fall to half strength; this is termed the half-life or $T_{1/2}$ and varies enormously from isotope to isotope (the radioactive form of element). For example, of those produced in an atomic reactor, Copper 64 has a half-life of 12.8 hr whilst nickel 59 has a 750 000 years half-life.

The first pollution hazard, therefore, is to be present where radiation leakage is occurring — this is relatively easy to avoid. The second hazard is where long continuous doses are endured at low levels and this is more of a problem. Atomic power stations, ships and nuclear detonations each produce radiation and only the latter is a critically dangerous source of pollution. However, large losses have occurred in power stations, so a constant need for monitoring is vital.

It is hard to believe, but in 1970 it was learned that the U.S. Atomic Energy Commission had in an underground store, some 20 x 10^6 gallons of radioactive waste much of which has half-lives measured in hundreds of thousands of years! Some isotopes are particularly dangerous. Strontium-90, for instance, accumulates in our bones encouraging cancer. Nuclear device testing in the early 60s did much to raise the normal background level.

Each radiation presents a different hazard, so the unit of strength is based on the biological effect it produces. This unit is the Roentgen equivalent man or rem for short. Normal background levels are around 0.1 rem per annum. Small doses greater than this can cause later-appearing symptoms. Large doses (hundreds of rem) will produce fever and digestive upsets that, if overcome, will lead to tumours and certain death at some stage. It is for these reasons that there is so much opposition to nuclear testing in the open atmosphere.

Alpha particles penetrate the least and are easily shielded or absorbed; Beta rays have the largest range, but gamma are the most penetrating. The relative quantities of each emitted depends upon the isotope.

The simplest detector of radioactivity dosages is the personnel-monitor worn on the lapel. This consists of a piece of photographic film half of which is shielded by a layer of absorber such as lead or aluminium. These cannot be read without processing.

Radioactive particles cause ionization and this is the principle used in the general purpose ionization detector shown in Fig. 14. Each RA particle entering the chamber ionizes the gas (air, argon, etc.) producing a current pulse that is amplified. The process is random, so a series of noise pulses are counted and averaged over a chosen time-period to be displayed on a meter or used to drive a loud-speaker unit. Certain filling gases have an amplification factor of a million. These, if used, enhance the sensitivity.

The Geiger-Muller tube is of the ionizing type and is typified by a characteristic that provides constant pulse sizes regardless of particle type. Many variations exist, depending on the shape and the operating voltage, but all are most inefficient using only 1% of the radiation passing through to provide an output signal.

Another disadvantage of ionization cells is that time is essential (0.1-0.5 millisec) for the anode to become sheathed by charge in readiness for the next particle event.

A superior, but more expensive, method for detecting RA is the scintillometer. Referring to Fig. 15 the incoming particle enters the crystal (of

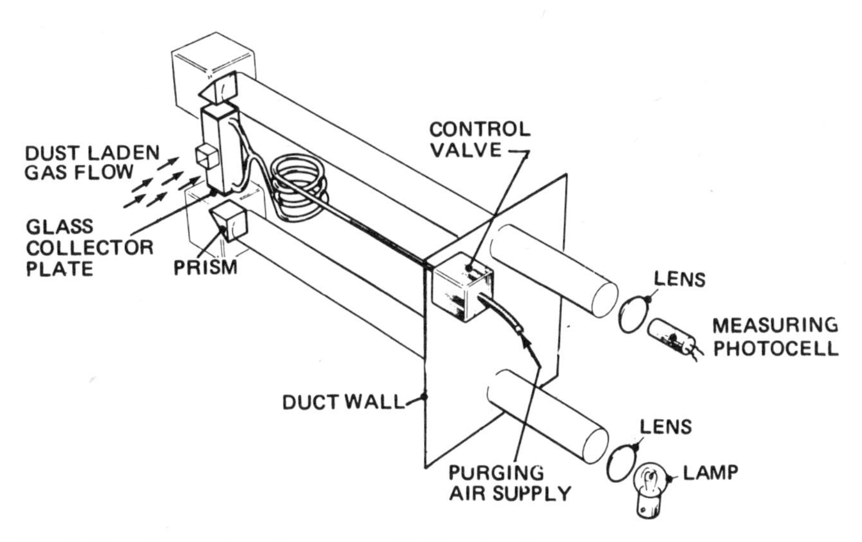

Fig. 13. Flue dust monitor, developed at CERL and marketed by Kent, uses two horizontal mirrors, the lower collecting dust as it falls.

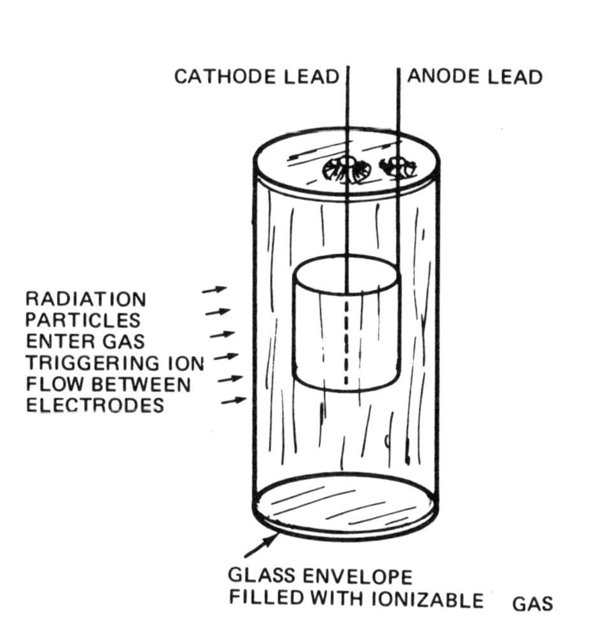

Fig. 14. Simple ionization chamber detects nuclear radiation particles.

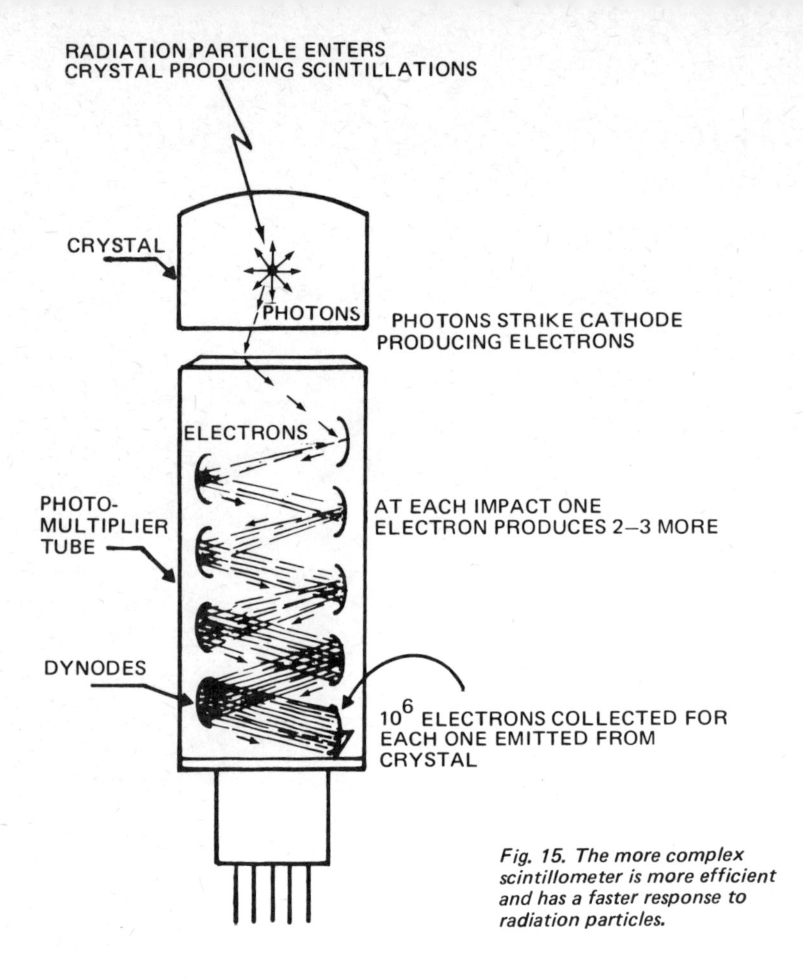

Fig. 15. The more complex scintillometer is more efficient and has a faster response to radiation particles.

stilbene or sodium iodide) where it releases photons that scintillate at visible radiation wavelengths. This energy conversion process is reasonably efficient and, furthermore, amplification of light can be had with extremely low-noise addition by the use of a photo multiplier as is shown in the figure. The time delay of scintillometers can be as small as 0.01 µs so more particles can be detected.

Effective use of these detectors involves the use of pulse processors to discriminate between coincident pulses, to produce averaged rates, and special needs such as pulse height discriminators for the detection of the form of radiation as well as its strength.

FURTHER READING

"Modern Methods of Chemical Analysis", R. L. Pecsok and L. Donald Shields, Wiley, 1968, London.

"Basic Gas Chromatography", H. M. McNair and E. J. Bonelli, Varian Aerograph, 1968, California.

"Practical Manual of Gas Chromatography", J. Tranchant, Elsevier, 1969, Amsterdam.

"Correlation Techniques Applied to Gas Chromatography", G. C. Moss and K. R. Godfrey, Trans Inst. Meas. Control London, 1972, 5, 351-353.

"Measurement of Oxygen Content in Gases", F. Tipping, as above, 1970, 3, 145-152.

"pH Facts – The Glass Electrode; The Industrial Scene", W. Thompson and E. Gill, Kent Technical Review, 1972, 7, 16-22.

"Variosens – Optronic Instrument for Measurement of Turbidity . . .", Impulsphysik GmbH, Hamburg.

"The Transmissometer", G. W. Oddie, Weather, 1968, November.

"The Continuous Monitoring of Particulate Emissions", D. H. Lucas, W. L. Snowsill and P. A. E. Crosse. Trans Inst. Meas. Control, (London), 1972, 5, 0-21.

"Neutron Irradiation and Activation Analysis", D. Taylor, Newnes, 1964, London.

"Radioactivation Analysis", H. J. M. Bowen and D. Gibbons, Oxford University Press, 1963, Oxford.

"Radioisotope Techniques for Measurement and Control of Industrial Pollution", J. F. Cameron, Trans. Inst. Meas. Control, 1971, 4, 303-307.

"Principles of Instrumental Analysis", D. A. Skoog and D. M. West, Holt, Rinehart and Winston, 1971, New York.

"Practical Applications for Infra-Red Technique", R. Vanzetti, Wiley, 1972, New York.

"Principles of Process Analysers", D. J. Huskins, Adam Hilger, 1979, Bristol.

"Particle Size Measurement", T. Allen, Chapman and Hall, 1975, London.

"Handbook of Air Pollution Analysis", Chapman and Hall, 1977, London.

CHAPTER 13
MEASUREMENT DIFFICULTIES, INSTRUMENT INFORMATION

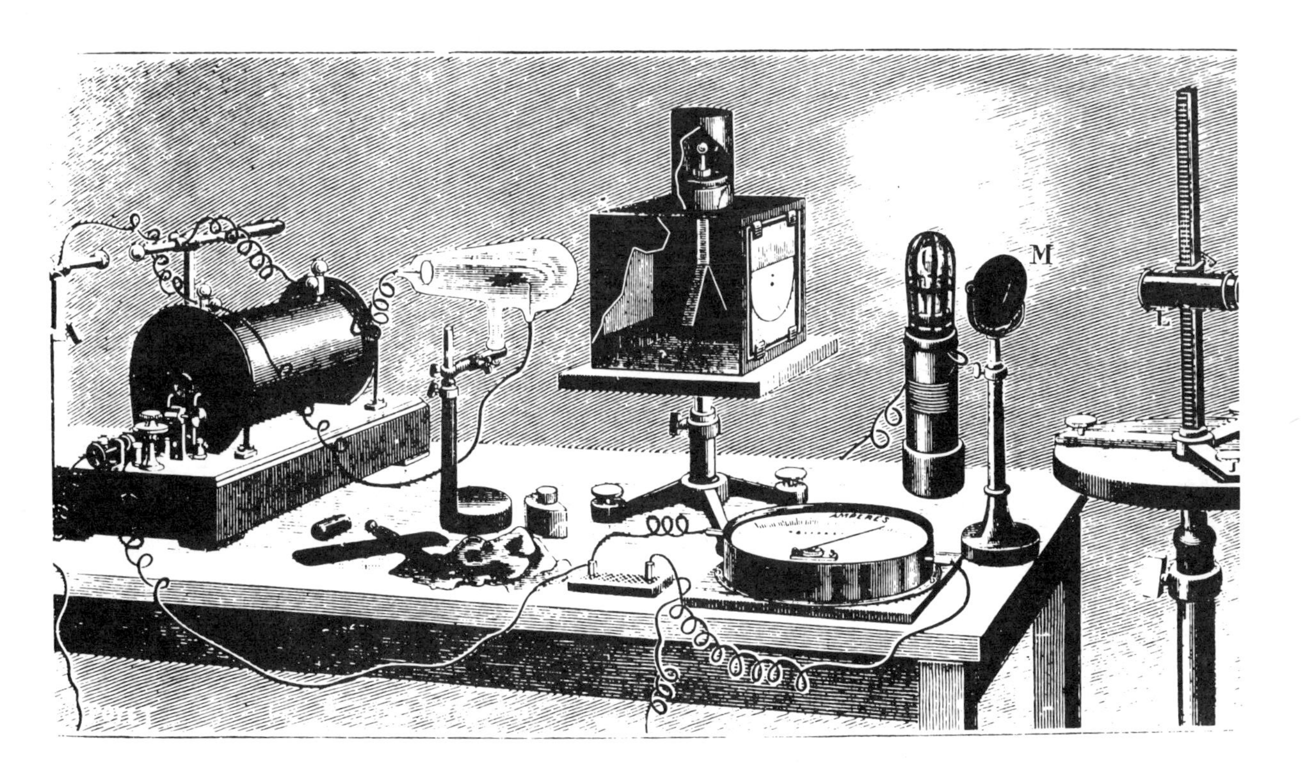

Discovery of new physical principles requires the invention of new measuring apparatus which, in turn, leads to the discovery of more principles on which to build yet more instruments.

This array of apparatus is typical of hardware used in 1896 to investigate gaseous discharges and X-ray radiation. Out of these comparatively simple instruments grew the sophisticated systems used to monitor radiation from radiographic and nucleonic equipment.

Development of transducers is a process of steady evolution.

We have looked at many ways by which the common physical variables are sensed and transduced into equivalent electric signals. The methods that we have described are well developed and are reliable and often extremely precise. Because of this it might be thought that there is a sensor available to monitor any possible variable and that the sensor design aspect of measurement systems engineering is now simply a case of applying commercially available devices in a routine manner. This is not the case — many highly desirable sensors are still not developed sufficiently to use, some have not even been realised; the need for inexpensive small transducers is paramount.

The task of the transducer designer continually includes the creation of new sensors (in-use tyre wear, leaf area increase rate on a tree, grass height monitor for automatic lawn-mowing, an education effectiveness monitor, an elasticity modulus meter for structural foundation bore-holes — these are all current problems awaiting solutions).

The task is not always to provide a parts per million precision. As illustrations of these difficulties, two

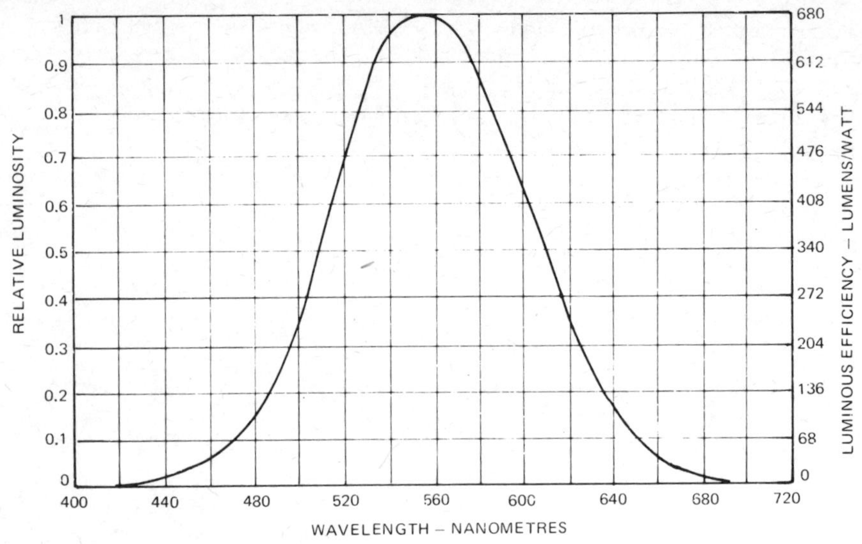

Fig. 1. Response curve of 'standard eye'.

areas of measurements are now outlined: the first being an old discipline where progress has been slow — photometry — and the other, viscosity, a relatively new automatic measurement need.

PHOTOMETRY

Photometry is the science of measuring visible radiation in relation to its ability to produce visual sensation. (Sometimes the definition is extended to include radiation bordering on the visible, namely, infrared and the ultraviolet regions). This is a psychophysical parameter, for determinations are related to the human eye and what it appears to see. Therein lies the difficulty — a standard average eye response is needed to allow for the biological variation in eyesight between observers.

We have already dealt with a better way to measure a radiation variable when black body radiation was discussed. These radiometric measurements do not rely on psychophysical factors. Photometry had its birth in the days before the work of Planck when scientists wished to quantify the effects of light. They created their own standards of brightness and illumination intensity using firstly, standard size candles and later standardised flames in which the gas jet orifice, the gas pressure and composition were carefully specified so that different constructors obtained much the same standard. The response of the eye varies with wavelength and the individual, so attempts were made to define a standard observer as the average of a large number of people. In 1951 this became accepted reality. A graph showing the agreed response is given in Fig. 1.

It is not feasible to employ a person with such a standard eye response, so photo detectors are used in conjunction with fresh liquid-filled optical filters to modify the response accordingly. Coloured glass filters are avoided as these degrade with time. Detectors used include phototubes and selenium cells. But even with careful filter design, agreement with the standard response cannot be made accurately so a residual error curve is used to correct readings when extreme accuracy is required.

An alternative method to obtain the response disperses the broad band white light source separating out the colours. These are then passed through a shaped aperture allowing the correct amount of light through at each wavelength. The colours are then recombined. It can be seen that these procedures require a high skill content and a lot of tailoring to obtain accurate results. Where feasible, photometric measurements are made on a relative basis as the eye is more sensitive to differences in brightness than to the absolute level of radiation.

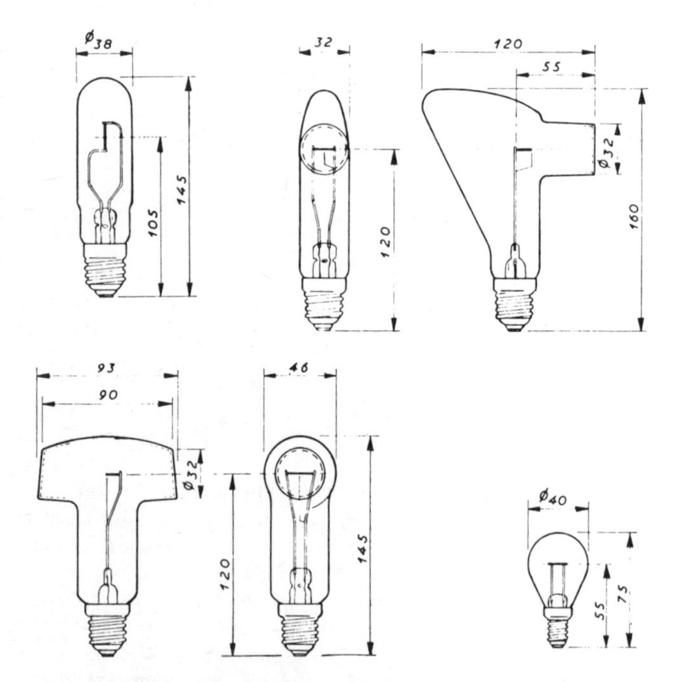

Fig. 2. Tungsten ribbon lamps especially designed to produce a uniformly radiating surface. The windows are ground plane-parallel from quartz or heat-resisting glass and must have their spectral energy transmission calibrated before manufacture. Size in mm.

Fig. 3. EEL apparatus for measuring the reflectivity of ore minerals (using microphotometry — an early method).

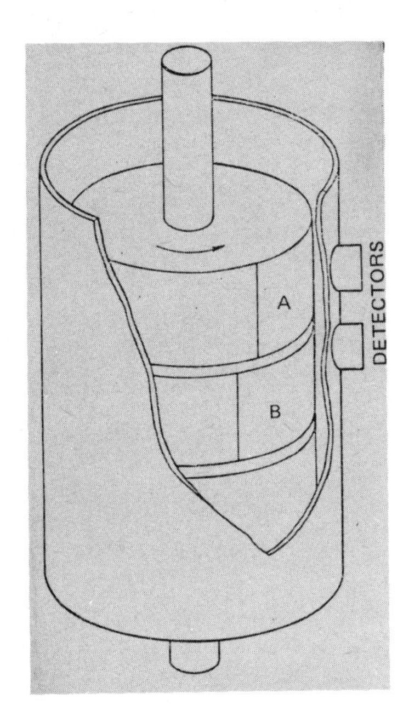

Fig. 4. Schematic of the Rosemount on-line rotating viscometer.

Radiometric measurements, not being based upon a subjective test, are usually to be preferred, but the extensive use of the photometric units in applications such as illumination engineering, in visually used optical instruments, in astronomy and in television prevents the total abandonment of the latter in favour of a radiometric alternative.

Luminous intensity, the 'candle-power' of a source, is now standardized as the candela (cd) a unit defined in terms of the luminous intensity of black-body radiation. To make life easier in day-to-day measurements, secondary standards — tungsten lamps of special design like that shown in Fig. 2 — are used. Luminous flux has units of lumens (lm) and relates to the flow of light fluxing away from the source. The candela is ideally a point source radiating into a spherical space around it. Allowing for this, a candela source emits 12.57 lumens of light flux. Illumination is the luminous flux falling on a surface and is expressed in lumens per square metre, units called lux (lx). The fourth unit, luminance, is similar to the lux except that this is the luminous intensity of an extended surface — not a point source. Its unit is the nit (nt) which is candelas per square metre.

Numerous other photometric units exist from the past (lambert, foot-candle, apostilb, phot, skot, talbot, metre-candle) but only the four, candela, lumen, lux and nit are to be used in future. As both radiometric and psychophysical measures relate to the same thing it is possible to equate luminous flux with radiated power once the standard eye is defined. In Fig. 1. the luminous efficiency of a source is represented on the right-hand axis. For example, a source radiating at 0.56 μm will produce 680 lm/watt.

The practical difficulties of creating a standard eye response, the primary standard black-body and secondary standard lamps are considerably greater than those encountered in many other forms of standard. The best precision attainable is barely better than one part in 1000. Similar problems exist in ultra-violet and infra-red 'photometry'.

Creating and maintaining photometric standards are seldom the tasks of the transducer user: acquaintance is more likely to be with photometer instruments of one kind or another. The microphotometer is an instrument incorporating photometric measurements with microscopy in order that the light magnitudes of very small areas can be measured. A common use of these is to determine the degree of blackening of photographic emulsion records from a spectrograph, a field of stars or interferometer fringes. In the automatic scanning photometer a narrow beam of light is passed through the emulsion to impinge on a photodetector. The photographic plate is then slowly moved across, recording the detector output with position.

Another application of the microphotometer is for measuring the reflectivity of mineral grains in order to investigate the composition and nature of an ore. One such equipment is shown in Fig. 3. Light of constant amplitude and wavelength composition is directed onto a cut and polished specimen held on the microscope stage. The reflected radiation is first viewed visually through the eyepiece to identify the area being measured; a thin slide containing a flat photocell is then inserted under the eyepiece to measure the light flux reflected. The cell is connected to the galvanometer via a preamplifier in order to amplify the microampere signals. Added sensitivity can be obtained with a split-photocell position-sensitive detector arrangement. This is clipped onto the scale of the galvanometer to amplify the movements of the light spot. The same apparatus can also be used to determine exposure times in microphotography. This example

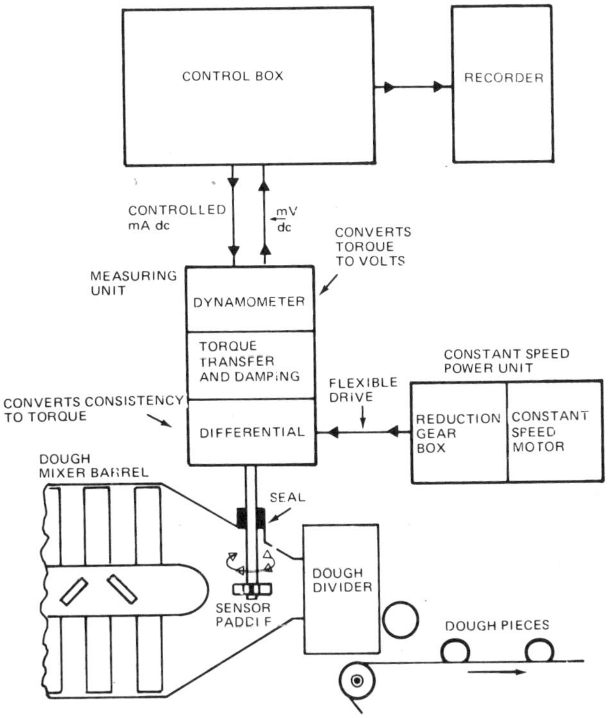

Fig. 5. Basic system components of consistency measuring instruments used in automated bread making.

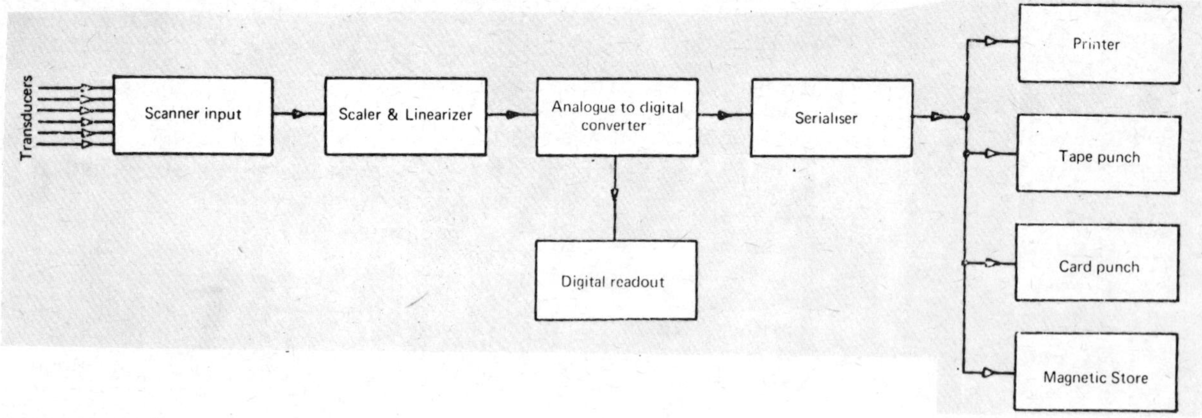

Fig. 6. Basic data-logger arrangement.

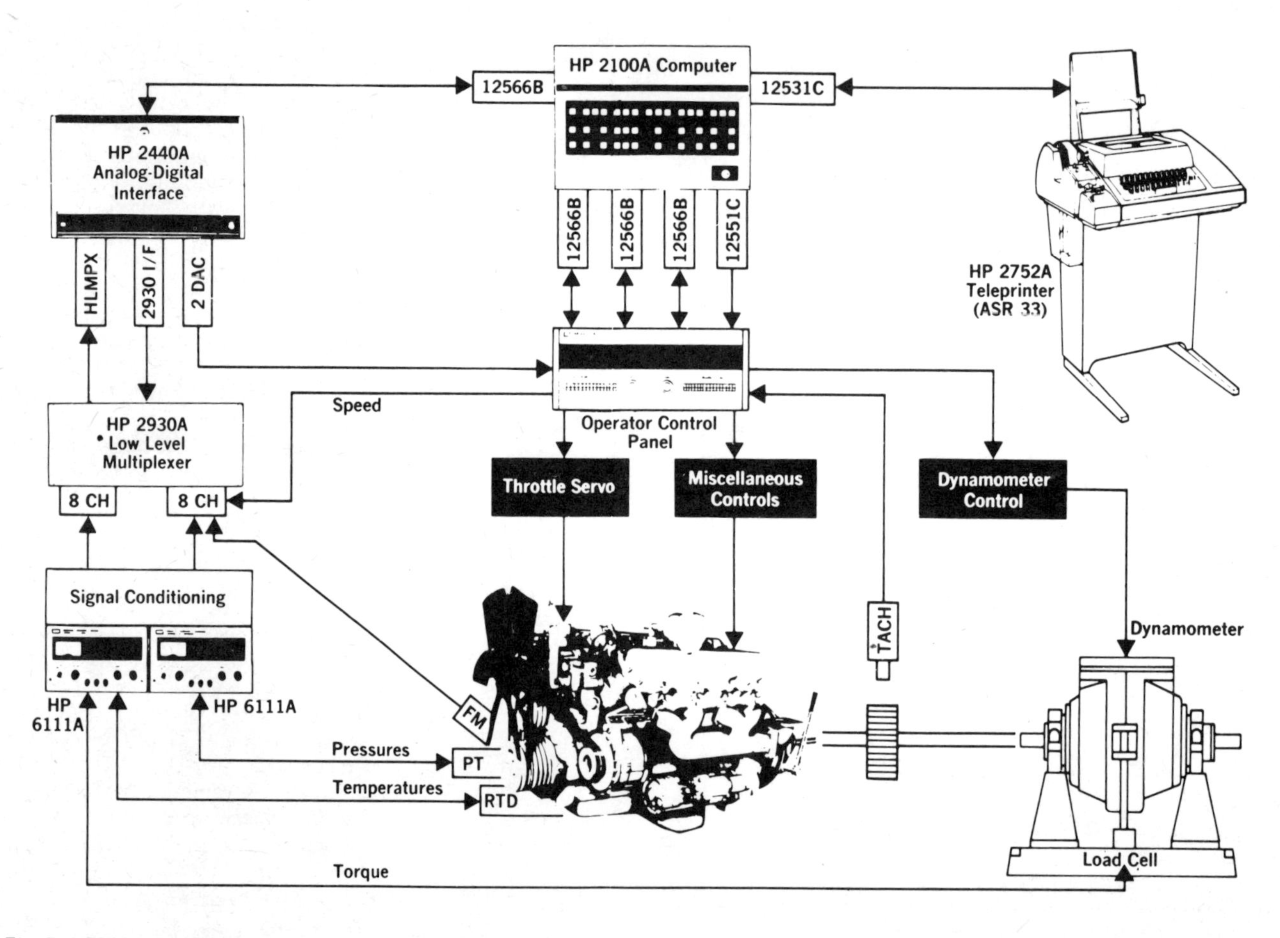

Fig. 7. HP9600 data acquisition and control system applied to engine testing.

shows how a seemingly simple measurement can become impossible as precision is increased, if the basic principles are inappropriate.

VISCOSITY

A fluid's resistance to the tendency to flow is called its viscosity. Instruments for measuring it are called viscometers or rheometers. Most viscometers make use of the viscous drag upon a rotating or oscillating cylinder turning in the fluid. In the unit shown diagrammatically in Fig. 4. the upper and lower cylinders are solidly connected and turn in the container filled with the fluid to be measured. The central bob, as it is called, is connected to the outer two by a taut ligament that allows it to lag behind the outer pair depending upon the viscous drag on its surface. External detectors sense some form of mark on the cylinders producing two pulse trains whose phase will depend upon the angular difference. Fluids fall generally into two classes of viscosity, those in which it is constant with shear rate (called Newtonian fluids) and those in which viscosity depends on shear rate (the non-Newtonian fluids). Examples of the latter are paints, creams, polymers and emulsions. In the illustrated viscometer, the pulse rate is a measure of shear rate so it can also be used to investigate the change

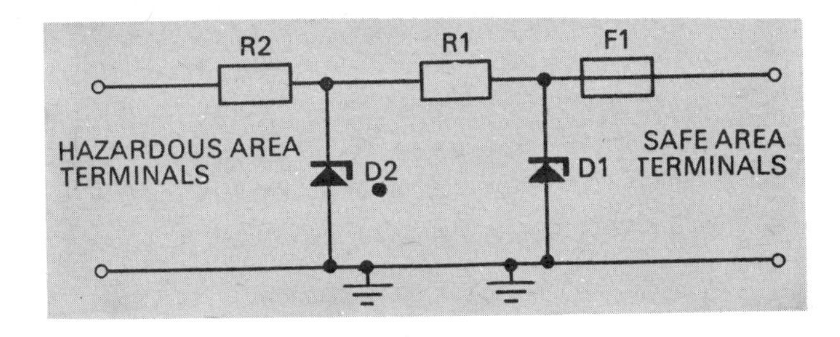

Fig. 8. Circuit used in a Zener barrier.

of viscosity with speed. (In viscosity measurements it is vital to control the temperature of the fluid as viscosity depends on it).

Until the recent adoption of the SI system of units, viscosity was measured in poise (after the 19th Century scientist Poiseuille). The SI unit to be used in future is the Pascal second (Pa s). One centi-poise is equal to 10^{-3} Pa s.

An unusual challenge arose in the on-line measurement of the viscosity (regarded as consistency in this case) of bread dough, sponge batter and pastry fat used in automatic baking production by the J. Lyons catering company in London. In the chosen solution, see Fig. 5. the highly viscous dough (5×10^{2} Pa s compared with 10^{-3} Pa s for water) is forced from a mixing stage out of a proportioning batching head. Resistance to shear within the dough is determined by the torque needed continuously to rotate a paddle immersed in the dough. A differential gear unit is used to sense the torque exerted by the paddle. This mechanical computing element provides a turning moment on the output shaft when a torque exists on the paddle being driven at constant speed. A strain-gauge load-cell converts the output torque into an electrical signal. It is appropriate in this industry to refer to the 'tightness' of 'slackness'

Fig. 9. Control console of Lucas automatic incinerator features indicating mimic panel.

Fig. 10. Human engineered control panel of computer-controlled water distribution station — controls are laid out in the mimic diagram.

of the dough, so they use a special off-line instrument called a Farinograph to calibrate the consistency meter in their own kind of units. It is able to handle products with viscosities in the range 500 to 0.1 Pa s when suitable paddles are used.

This example well illustrates the need for inventive designers in transducer development.

TRANSDUCERS AND PROCESSES

Automatic measurement and control of processes could not be realised if transducers did not exist. They are the essential communication link between the physical attribute of interest and the recorder or controller. The number of applications for sensors is infinitely large, ranging over disciplines such as medicine, power generation, marine survey, weapons, food-stuffs processing, oil refining, scientific research, pollution control, television and communications.

In many cases the need is for a large number of parameters to be monitored requiring the use of a diverse arrangement of sensors as we have seen exist. Sometimes they are of the same kind but at different locations, for instance, as in the structural testing of a ship at sea using strain gauges, or they may be of different kinds as in a process plant where temperatures, pressures, moisture content etc., must be monitored. When data from each are not needed continuously, they can be switched sequentially onto a common line feeding a single time-shared recorder — as shown in the data logger schematic of Fig. 6. When automatic data processing is used, a digital form of output is preferred — some transducers are designed to produce digital signals directly. For visual monitoring, analogue chart records are best. It is common practice to use both to satisfy both needs. An essential requirement in large-scale development is that each sensor delivers signals that are compatible and uniformly ranged to make best use of the dynamic range of the recorder. Many instrument manufacturers now list data acquisition systems as product lines: the Hewlett Packard 9600 system is shown in Fig. 7.

To assist the design of large, complex process-control systems, makers of transducers and controllers use a number of standardised transmission signal ranges, examples being the 4-20 mA dc and 0-10 mV dc systems, but these are not universally adhered to; recalcitrant manufacturers claiming that their particular output range is more satisfactory.

SAFETY MEASURES

Often sensors must be situated in hazardous areas where explosion could result from a spark or excessive heating of the sensor circuits. The obvious way to reduce this risk is to contain everything in flameproof enclosures of great strength. There are disadvantages to this method, namely, that the cost is high, and that testing and maintenance are made difficult, for the power must be shut-off when the enclosures are opened.

The alternative, and more recently adopted method, is known as intrinsic safety. Inflammable vapours require a specific level of energy to bring them to the flash point. Limiting the amount of power available from a sensor circuit under any condition will eliminate this risk. No enclosures are needed and the sensor can be adjusted with the power on.

Originally the concept was implemented by ensuring that the circuit could not draw or give up greater than a specified power in the event of a fault. (This amount is found by testing in a special test chamber filled with inflammable vapour). A marketed Philips pressure transducer system is maintained intrinsically safe by limiting the inductance to 33 mH (at the most) and the capacitance to 180 nF. It uses the 4-20 mA standard (representing 0-100% signal swing) to provide power from the quiescent 4 mA value. This method works but suffers from the need individually to check each circuit for safety. A plant having hundreds of transducers would take a considerable time to inspect.

A recent improvement (it was introduced around 1960) is to use a circuit placed on the boundary of the hazardous area and the non-hazardous area that can ensure that all electronic circuits connected after it in the unsafe zone are isolated with regard to high power levels. The Zener barrier does just this. Basically it uses a Zener diode connected across the two-wire line and a fuse in series on the unearthed safe side line. A power surge attempting to pass the barrier (perhaps as the result of a sensor or wiring fault or a fault at the supply end) and thus attempting to raise the voltage at the sensor, is diverted to earth by the diode. Faults of long duration that might destroy the Zener diode are eliminated by the fuse blowing. In practice, a Zener barrier uses a pair of diodes, (see Fig. 8.) to reduce the risk of losing protection by an open-circuit diode (that would not be detected). Zener barriers, complying to Standards, are available commercially as ready-made units. As yet, their use is not internationally accepted. For further details consult the article listed in the suggested reading.

CONTROL ROOMS

Today, extremely complex processes and large power plants can be operated by a staff of only two or three people. This is achieved by monitoring all essential parameters with transducers feeding their signals back to a control room. By making the controllers and recorders small enough and of standard shape it is possible to mount them together on a control console.

Well designed layouts are the result of considerable thought, for the operator must instinctively know what to do in an emergency. Two basic arrangements are as follows: The layout can contain a schematic or mimic diagram of the plant on which lights operate to show failure or correct action. The console of a computerised refuse disposal system is shown in Fig. 9. Alternatively, the controls themselves can be placed in the mimic layout covering the whole panel as shown in Fig. 10. Row after row of exactly similar knobs and dials are to be avoided if each has a different purpose. Often a multi-channel closed-circuit television monitor is incorporated to enable the operator to view the proceedings at selected places.

SOURCES OF INFORMATION AND SERVICES

Anyone who has had to select or make a sensor for a specific task will know the frustrations involved — does such a sensor exist on the market; where does one buy it; are the specifications realistic in light of current technological achievement? Often a sensor is built from scratch (at great cost) because the task of researching the literature and manufacturers' catalogues for information appears prohibitive. Because the sensor may be cheap does not imply easy purchasing. It is, therefore, appropriate to outline where assistance is available in this regard.

If the sensor is known to be in common useage, the relevant Standards Association leaflets, kept in most libraries, will describe what is accepted and attainable practice, for the Standards are reached by considering current *practice, not future hopes.* This study may also reveal specifications of the sensor that may have been overlooked. Manufacturers' data sheets also assist in understanding the pros and cons of a device, but read more than one maker's literature for it is not totally unknown for a manufacturer to be biased in favour of his own product!

If you are not fortunate enough to have a good technical literature library at your disposal, assistance may be available from places such as the Standards Laboratory (NPL in Britain, NML in Australia, NRC in Canada, NBS in the States, etc.) of the country concerned, and also from the government laboratories.

To help ensure that marketed instruments (and other products) are up to specification, each country has a national testing authority that inspects and registers testing laboratories who have the necessary facilities. In Britain the British Calibration Service, BCS. In Australia this is the National Association of Testing Authorities, NATA. Only laboratories maintaining the required standard of equipment and excellence of use can obtain registration. Consequently, a study of the index listing the certificated laboratories may reveal a manufacturer in the field of interest. These laboratories can test equipment for buyers as well as manufacturers, at a reasonable charge.

Specialized services exist in technological sophisticated countries. The association in Britain concerned with instruments is SIRA Institute. They operate a service called SIRAID on behalf of their numerous member companies. Enquiries regarding makers, suppliers and data of instruments are attended to promptly, free of charge. A phone call, letter or telegram asking for a list of potential suppliers of a named transducer will result in a letter by return. The service goes further, for return. The service goes further, for reasonable rates, to provide assistance in design and testing, consultation services, and for specific research into instrument problems. SIRAID is not confined to British residents; their address is South Hill, Chislehurst, London, U.K.

Australia has a peculiar problem due to the high content of imported equipment — this is the task of locating the agent representing the manufacturer. A simple way is to contact the respective Embassy or special office of each country, such as the British High Commission for British products. The staff are well informed and usually maintain an up-to-date library of products and trade journals supplied by air mail.

Sources of aid often disregarded are Universities and Technical Colleges. It is a common feature in many countries that industry shuns consulting the academics. It is true that some tertiary research is esoteric and not relevant to day-to-day problems, but in general these institutions contain a wealth of free information. The staff are usually willing to assist for this is one of their roles in society.

If the interest is research orientated, it might pay to subscribe to one of the personal abstracting services such as ISI (Institute of Scientific Information). For a modest fee they will regularly supply selected abstracts, from numerous journals, in which the title includes key words chosen by the subscriber. ISI also publish the Science Citation Index, a series of volumes in which authors referred to by writers of articles are listed. This rapidly assists the reader to find out who else is working in the same field, and thus find other information on a technique.

We have now reached the end of this discussion. It has been the intention in this book to promote awareness of the many techniques and their limitations. No universal sensor appears to exist, so the task of selection will continue to be a matter of careful consideration. There will, however, always be a universal need for sensors to convert the real-world phenomena into a language suited to our automatic communication methods.

FURTHER READING

"Physical Photometry" — B.H. Crawford. Notes on Applied Science No. 29, National Physical Laboratory, H.M. Stationery Office, London, 1962.

"Luminance — Light Units — Illumination" — Mullard Australia Leaflet No. MTP 1062.

"Photometry" — J.W.T. Walsh, Constable, London, 1958.

"An In-Line Consistency Meter for Dough-Like Materials" — A.T.S. Babb and C.B. Casson, Measurement and Control, 1970, 3, T173 — 180.

"Functional Modularity Helps Designer and User of New Measurement and Control Subsystem" — J.M. Kasson, Hewlett-Packard Jnl. 1972, 23, 12, 13-19.

"The Nerves and Brains of Modern Industry" — L. Finkelstein, "Improvement", 1972, George Kent Group, Luton, England. 22-23.

"Safety Barriers Around the World" — D.J. Gaunt and A.T. Mead Kent Technical Jnl. 1972, No. 7. May, 28-31.

INDEX

W

X, Y, Z